WAFL 2011

Proceedings of the 5th International Conference on the Assessment of Animal Welfare at Farm and Group Level

WAFL 2011

Campbell Centre for the Study of Animal Welfare
University of Guelph
Guelph, Ontario, Canada
8-11 August 2011

edited by:
Tina Widowski, Penny Lawlis and Kimberly Sheppard

ISBN: 978-90-8686-180-4
e-ISBN: 978-90-8686-738-7
DOI: 10.3921/978-90-8686-738-7

First published, 2011

Wageningen Academic Publishers
P.O. Box 220
6700 AE Wageningen
The Netherlands
www.WageningenAcademic.com
copyright@WageningenAcademic.com

Welcome

The Campbell Centre for the Study of Animal Welfare (CCSAW) at the University of Guelph was established in 1989 and has over twenty years of experience in promoting improved animal welfare through research and teaching and through partnerships with animal industries and governments in Canada and throughout the world.

We are pleased to have the opportunity to host the 5th International Conference on the Assessment of Animal Welfare at Farm and Group Level (WAFL). The first WAFL – at that time a Workshop – took place when the science of animal welfare assessment was in its infancy. Now, twelve years later, on-farm animal welfare assessment and audits have evolved to become part of every day life for many of those involved in animal agriculture. With over a decade of experience we can now begin reflecting on lessons learned from existing animal welfare schemes, we can examine the broader aspects of animal welfare assessments – environmental and social aspects – and we can consider the impacts and benefits of animal welfare assessments on both the end-users and the animals.

In keeping with the spirit of the first WAFL, we will also explore new and emerging methods for animal welfare assessment, focusing on the validity, repeatability and feasibility of different animal welfare measures.

We would like to thank all of the people, including our students, who have made this conference possible. We would also like to acknowledge the tremendous interest and financial support that we have received from food animal industries, government and professional associations.

Welcome to Canada. We hope that you enjoy the conference.

Tina Widowski, Cate Dewey, Penny Lawlis and Kimberly Sheppard

Acknowledgements

Scientific committee

Tina Widowski (Chair)	University of Guelph
Penny Lawlis (Co-Chair)	Ontario Ministry of Agriculture Food and Rural Affairs, Canada
Lotta Berg	Swedish University of Agricultural Sciences
Renée Bergeron	Université de Guelph–Campus d'Alfred
Cate Dewey	Ontario Veterinary College, University of Guelph
Anne Marie de Passillé	Agriculture and Agri-food Canada
Jeff Rushen	Agriculture and Agri-food Canada
Suzanne Millman	Iowa State University, USA
Frank Tuyttens	Institute for Agricultural and Fisheries Research, Belgium
Elsa Vasseur	University of British Columbia, Canada
Isabelle Veissier	INRA, France

Organizing committee

Kimberly Sheppard (Chair)	Campbell Centre for the Study of Animal Welfare, University of Guelph
Tina Widowski (Co-Chair)	University of Guelph
Cate Dewey	University of Guelph
Trevor Devries	University of Guelph – Kemptville Campus
Derek Haley	University of Guelph
Penny Lawlis	Ontario Ministry of Agriculture Food and Rural Affairs, Canada
Stephanie Torrey	Agriculture and Agri-Food Canada
Patricia Turner	University of Guelph
Jackie Wepruk	National Farm Animal Care Council, Canada

Referees

Lotta Berg
Renée Bergeron
Anne Marie de Passillé
Trevor Devries
Cathy Dwyer
Frances Flowers
Diane Frank
Jenny Gibbons
Neville Gregory
Stefan Gunnarsson
Derek Haley
Jan Hultgren
Anna Butters-Johnson
Ute Knierim
Penny Lawlis
Christine Leeb
Joop Lensink
Lena Lidfors
Jenny Loberg
Georgia Mason
Kristina Merkies
Marie-Christine Meunier-Salaün
Suzanne Millman
Lena Munksgaard
Jeff Rushen
Jen Tin Sørensen
Joe Stookey
Caroline Stull
Mhairi Sutherland
Stephanie Torrey
Anita Tucker
Frank Tuyttens
Elsa Vasseur
Antonio Velarde
Nina Vonkeyserlingk
Dan Weary
Becky Whay
Tina Widowski
Christoph Winckler
Jenny Yngvesson

Thank you to our sponsors

Diamond Sponsor

Platinum Sponsor

Gold Sponsors

Canadian Food Inspection Agency | Agence canadienne d'inspection des aliments

Canadian Pork Council
Conseil canadien du porc

IceRobotics

Dairy Farmers of Canada

Les Producteurs laitiers du Canada

together with

Silver Sponsors

MAPLE
LEAF

Gray Ridge
EGG FARMS

BURNBRAE
FARMS • FERMES

pork
checkoff

Bronze Sponsors

Contributor Sponsors

Conestoga Meat Packers

Pfizer Canada Inc.

Turkey Farmers of Canada

Turkey Farmers of Ontario

General information

Conference venue

The WAFL conference will be held at the University of Guelph. It will begin with a welcome reception in the Science Complex Atrium. Oral presentations will be in Rozanski Hall, and poster presentations will be in Peter Clark Hall (Level 0 of the University Centre). All locations can be found on the map below.

For further information regarding locations on campus please visit the University of Guelph website at http://www.uoguelph.ca/

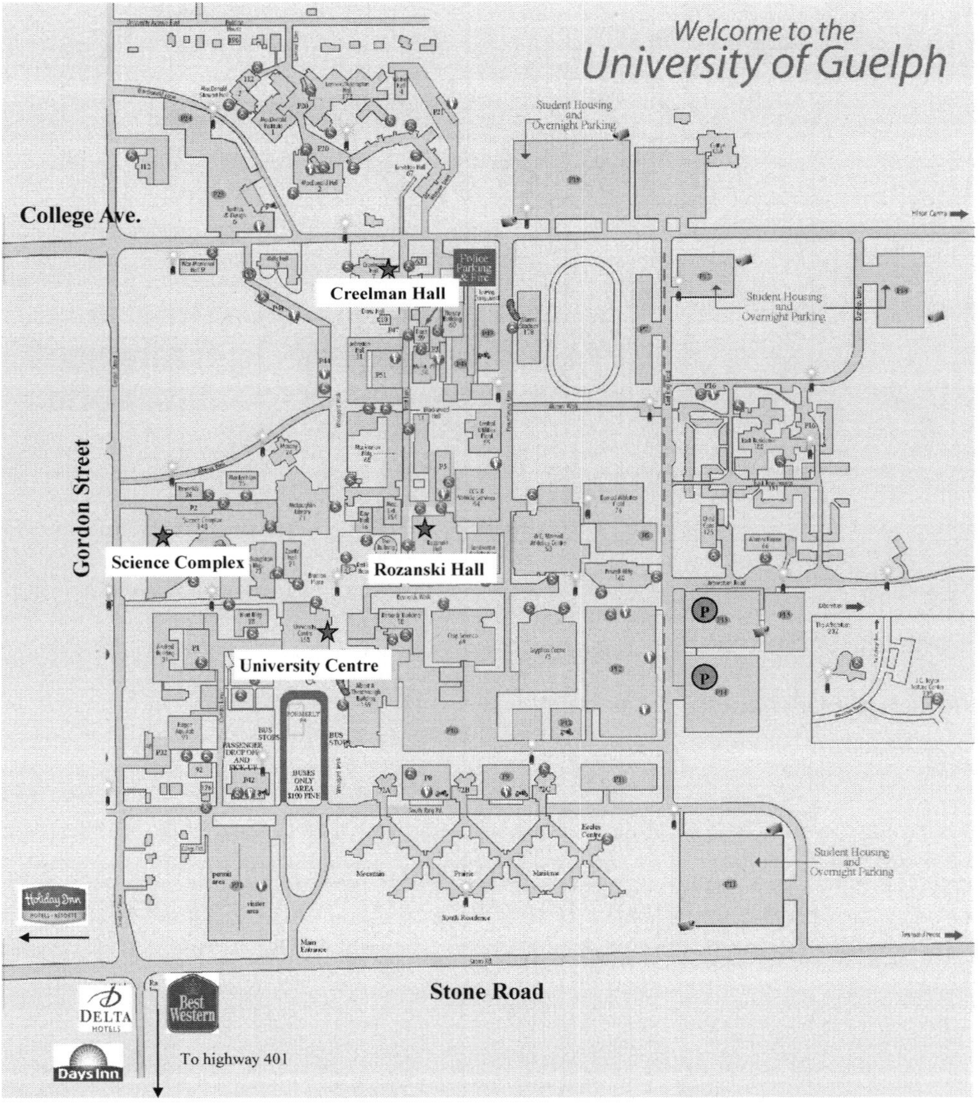

Official language

Official language of the meeting is English.

Registration and information desk

Monday August 8th	14:00 – 17:00	(Rozanski Concourse)
Monday August 8th	18:00 – 20:00	(Science Complex Atrium)
Tuesday August 7th	7:30 – 11:00	(Rozanski Concourse)
Wednesday August 10th	7:00 – 9:00	(Rozanski Concourse)

Name badges

Your name badge is your admission to the venues, scientific sessions, poster sessions, lunch and coffee breaks. Please wear it at all times during the WAFL Conference.

Poster and exhibition area

The poster and exhibition area will be in Peter Clark Hall, located in the University Centre, Level 0. Please visit the poster sessions where we will also be holding a wine and cheese reception on Tuesday August 9th, and lunch on Wednesday August 10th.

Internet access

Wireless internet is available. Your username and password can be found in your nametag.

Certificate of attendance

Your certificate of attendance is inserted in this book.

Coffee breaks

All coffee breaks will be held in the Rozanski concourse

Lunches

Lunches will be served in Creelman Hall on Tuesday August 9th, and Peter Clark Hall on Wednesday August 10th.

Welcome reception, August 8th, 18:30 – 20:30

The welcome reception will be held in the Science Complex Atrium. Hors d'oeuvres will be served and there will be a cash bar. You will be provided one drink ticket for this event.

Banquet, August 10th, 18:30

The conference banquet will be held in Creelman Hall.

Banking service, currency

The Canadian Dollar (CAD) is the official currency in Canada. Global Currency Services is located at 1027 Gordon Street, Guelph, Ontario. Major credit cards are accepted at most hotels, shops and restaurants.

Shopping in Guelph

Most stores in downtown Guelph are open from 9:00-17:00, Monday-Saturday, with many having extended hours until 21:00. Grocery stores usually have longer opening hours and many of the larger drug stores are open until midnight.

Emergency calls

You should call 911 if anything happens that requires an ambulance, the police, or the fire department. This number may be called from a fixed or mobile phone.

The number for University of Guelph campus police is x52000 from any University of Guelph phone.

Programme at a glance

Monday August 8th	
14:00 – 17:00	Registration open (Rozanski Concourse)
18:00 – 20:00	Registration open (Science Complex Atrium)
18:30 – 20:30	Welcome Reception (Science Complex Atrium)
Tuesday August 9th	
7:30 – 11:00	Registration open (Rozanski Concourse)
8:30	Welcome and Opening Remarks
8:55 – 10:15	**Session 1**: Farm animal welfare assessment and sustainability – Connecting social and environmental aspects of measuring welfare
10:15 – 10:30	Free papers
10:30	Coffee Break
11:00 – 12:15	Free papers and discussion
12:15	Lunch (Creelman Hall)
13:30 – 14:50	**Session 2**: Lessons learned from new and existing welfare assessment schemes
14:50 – 15:05	Free papers
15:05	Coffee Break
15:35 – 17:10	Free papers and discussion
17:30 – 19:00	Poster Session, Wine and Cheese (Peter Clark Hall)
Wednesday August 10th	
7:00 – 9:00	Registration open (Rozanski Concourse)
8:00 – 9:20	**Session 3**: Welfare assessment during transport, euthanasia and slaughter
9:20 – 10:05	Free Papers
10:05	Coffee Break
10:35 – 12:00	Free papers and discussion
12:00	Lunch/Poster Session (Peter Clark Hall)
13:30 – 14:50	**Session 4**: Our experience beyond the farm – animal welfare assessment for zoos, laboratories and animal shelters
14:50 – 15:35	Free Papers
15:35	Coffee Break
16:05 – 17:00	Free papers and discussion
18:30	Banquet (Creelman Hall)
Thursday August 11th	
8:00 – 9:35	**Session 5**: New and emerging methods for animal welfare assessment
9:35	Coffee Break
10:05 – 12:10	**Session 6**: Working with end-users of animal welfare schemes
12:10	Closing remarks

Animal Welfare Education Centres – Poster summaries

The American Veterinary Medical Association

AVMA Animal Welfare Committee: educating veterinarians and the public regarding best practices for animal welfare

The American Veterinary Medical Association (AVMA) represents a diverse group of over 80,000 veterinarians worldwide. A strategic priority for the AVMA is to be a leading advocate and authoritative resource for animal welfare both for the veterinary community and the public at large. The AVMA also believes that veterinarians must be willing and capable of accepting leadership roles as authorities on animal welfare. To achieve this goal, the AVMA's Animal Welfare Division and 18-member Animal Welfare Committee (AWC) are charged with proactively seeking and identifying animal welfare concerns and opportunities, critically evaluating information and stakeholders, and determining which activities and actions might be most appropriate to address concerns. In cases where no or insufficient information exists, the Committee must develop a plan to effectively address the knowledge gap in animal welfare. To educate veterinarians regarding refinements in veterinary medical care, the AWC has developed a set of policy statements that govern many aspects of animal use by humans, as sporting animals, food sources, research subjects, and companion animals. These policies are regularly reviewed and updated by the AWC to incorporate up-to-date evidence for enhancing animal welfare. The AWC also sponsors national and international educational sessions and symposia related to animal welfare, as well as several other high profile projects that significantly impact animal care and use globally, including the AVMA Guidelines on Euthanasia and the AVMA Model Animal Welfare Curriculum. This poster will address the role of the AVMA AWC in educating veterinarians and the public about animal welfare refinements.

Daniel Marsman[1], Gail Golab[2], Julie Dinnage[3], Patricia V. Turner[4], [1]Procter & Gamble, [2]AVMA Animal Welfare Division, Schaumberg, IL, [3]Association of Shelter Veterinarians, Scottsdale, AZ, [4]Dept of Pathobiology, University of Guelph, Guelph, ON

The Animal Transportation Association

The Animal Transportation Association (ATA) was organized in 1976 by industry leaders, government officials, and humane association representatives in response to concerns over the transport of animals. The ATA believes that all animals are a vital part of our world environment and are to be humanely treated at all times. The ATA understands the need for research, education and regulation of animal transportation and it is the policy of ATA to develop and promote, in collaboration with the industry, the best means of accomplishing these objectives. The ATA *Education Committee* has developed several tools to provide education on the humane transport of animals. These tools include *Best Practices During Transportation* for livestock, zoologicals, horses and lab animals from which transporters are encouraged to develop animal welfare assessment programs. *Factsheets* are available on the website that cover a wide variety of transportation issues and advice on how to insure that animals are transported under optimal conditions. *Webinars* are offered throughout the year, providing access to industry leaders from around the world, internationally recognized researchers and government officials providing timely information and education on the welfare of animals during transport. The ATA brings the transportation industry together to disseminate the latest information on the transport of animals at its annual conference held in different areas of the world each year. The organization further encourages uniform and effective international regulations and humane handling of live animals by continuing to work with government bodies on policy development regarding the welfare of animals during transport.

Jennifer A. Woods[1,2], Kelly Wheeler[3,4], Jeff Segers[5,6], [1]J. Woods Livestock Services, Blackie, Alberta, Canada, [2]Animal Transportation Association, Board of Directors/Livestock Chair/Education Committee, [3]Aviagen, Huntsville, Alabama, [4]Animal Transportation Association, President/ Education Committee/Presenting Author, [5]Skyfast, Brussels, Belgium, [6]Animal Transportation Association, Board of Directors/Regulatory Committee, rturner@drohanmgmt.com

British Columbia Society for the Prevention of Cruelty to Animals

Use of evidence-based research in education and outreach: the BC SPCA approach to farm animal welfare

The BC SPCA is the largest animal welfare organization of its kind in North America, and a Canadian leader in evidence-based education and outreach initiatives that improve the welfare of farm animals. The BC SPCA's vision is *to inspire and mobilize society to create a world in which all animals enjoy as a minimum, the Five Freedoms*. Farm animal welfare initiatives include:

- *SPCA Certified*: a farm certification and food-labelling program designed to provide farmers with product branding that allows them to market to welfare-conscious consumers and receive a price that rewards them for their higher costs of production. SPCA Certified farm animals are not confined to cages (laying hens) or gestation stalls (sows); subjected to unnecessary painful procedures (e.g. tail docking of dairy cattle), or fed antibiotics for growth enhancement; and their environments are designed to promote natural behaviours and healthy social interactions. In short, *SPCA Certified* farm animals live a higher quality of life.
- Two online research extension tools for producers: *FarmSense* e-newsletter and the *Resources for Farmers* webpage at spca.bc.ca/farm.
- Participation in the National Farm Animal Care Council (NFACC) and its projects, including development of the Canadian Codes of Practice for the Care and Handling of Farm Animals, and as the lead representative of the Canadian Federation of Humane Societies (CFHS).
- Youth education on farm animal welfare through *Cluck!*, a school curriculum unit on egg-laying hens, and *Bark!*, the BC SPCA's magazine for kids.

Alyssa Bell Stoneman, Brandy R. Street, Geoff Urton, BC SPCA Farm Animal Welfare Department, Vancouver, B.C.

Canadian Association for Laboratory Animal Medicine

CALAM: promoting research animal welfare coast-to-coast

The Canadian Association for Laboratory Animal Medicine/L'Association Canadienne de la Médicine des animaux de laboratoire (CALAM/ACMAL) was founded in 1982 and is the national organization that represents the interests of Canadian laboratory animal veterinarians working to support the humane care and use of animals used in research, teaching, and testing. The vision of the organization is to be recognised and respected as leaders in laboratory animal welfare. The central document to the CALAM/ACMAL vision is the *Standards of Veterinary Care*, which was last updated in 2007. The document emphasizes that CALAM/ACMAL and its individual members have a responsibility to provide leadership in developing best practices for the humane care and use of animals in research, teaching, testing and production, with due consideration of the 3 Rs: replacement of animals used, when possible; reduction of numbers of animals used; and refinement of techniques and procedures employed. CALAM/ACMAL recognizes that the well-being and welfare of animals used in research, teaching, and testing are the main focus for all laboratory animal veterinary roles and responsibilities. For laboratory animal veterinarians, animal welfare includes physical and behavioural aspects of an animal's condition, evaluated in terms of environmental comfort, freedom from pain and distress, and provision of appropriate social interactions. The organization promotes a number of educational tools, symposia, and fellowships to veterinary students and veterinarians to facilitate the exchange of knowledge and harmonization of standards of veterinary care for Canadian laboratory animal veterinarians. This poster will describe CALAM/ACMAL's role in promoting research animal welfare.

Patricia V. Turner[1], Andrew Winterborn[2], [1]Dept of Pathobiology, University of Guelph, Guelph, ON, [2]Animal Care Services, Queen's University, Kingston, ON

Canadian Council on Animal Care

Considerations for the welfare of farm animals used in science

The *Canadian Council on Animal Care guidelines on: the care and use of farm animals in research, teaching and testing* (2009) provides standards for housing, husbandry and use of farm animal species maintained in Canadian institutions (academic, government and private) for any scientific purpose. The document is intended to assist those involved in the care and use of the animals, keeping in mind the wide variety of areas in which farm animals are used: from non-invasive studies of feed to vaccine efficacy testing, and as human disease models or models for medical device testing. Studies with goals related to animal production under commercial conditions presented further challenges, since our guidelines aim to provide a higher standard of animal welfare for those animals used in science. However, the document recognizes that requirements for housing and husbandry should be determined in accordance with the particular scientific goal. In setting the guidelines, we used the same approach taken with the development of other CCAC guidelines documents, i.e. the focus was placed on the animals themselves. We asked the question 'what do the animals need in order to be normal?', and then considered how these needs could be addressed in light of the particular scientific goal. In doing so, we differentiated between improvements to an animal's environment that would enable it to express behaviours that it is highly motivated to perform, versus enrichment that would further enhance the animal's quality of life. In order to determine how well an animal is coping in its environment, it is important to carry out welfare assessments. This is encouraged for all animal species used for scientific purposes, including farm animals. In the case of invasive studies, welfare assessment provides a valuable tool for determining the endpoints of studies, particularly where pain and distress may be involved.

G. Griffin, J. Dale, Canadian Council on Animal Care, 1510-130 Albert St., Ottawa, Ontario, K1P 5G4

Canadian Veterinary Medical Association

Effecting change in animal welfare at a national level – the role of the Canadian Veterinary Medical Association

Refining animal welfare practices is a key priority for the Canadian Veterinary Medical Association. The organization is a national science and ethics based resource for veterinarians, veterinary students, provincial and federal governments, and the public regarding animal welfare standards for all species. Through the actions of its Animal Welfare Committee (AWC), the CVMA seeks to educate members and others regarding acceptable practices for animal care and use, leading to enhancement of animal well-being. The CVMA's AWC achieves their goals in a number of ways. Leadership and active advocacy for enhancing animal welfare is demonstrated through development of position statements that consider current knowledge and scientific evidence to refine treatment and care of animals. These position statements are developed with member consultation and advice from pertinent stakeholder groups. In addition, the CVMA's AWC develops and disseminates educational tools, such as posters on various topics, for veterinarians, students, and the public, which are strategically geared to create shifts in thinking about currently accepted practices; maintains a roster of spokespersons to address national animal welfare issues, they publish peer-reviewed papers on current animal welfare topics; and develops codes of management and husbandry practice for species not covered by national producer organizations. The CVMA takes an active advocacy and lobbying role at a national level with a number of nongovernmental and governmental organizations and committees to effect policy changes in accepted best practices for animal welfare. This poster will explore the CVMA's role in effecting enhancements for animal welfare in Canada.

Patricia V. Turner[1], Terry L. Whiting[2], Warren Skippon[3], [1]Dept of Pathobiology, University of Guelph, Guelph, ON, [2]Office of the Chief Veterinarian, Winnipeg, MB, [3]CVMA, Ottawa, ON

Danish Centre for Animal Welfare

The Danish Centre for Animal Welfare (DCAW) was initiated in January 2010 and is located at the Danish Veterinary and Food Administation. The centre is a collaboration between animal welfare experts from the Danish Veterinary and Food Administration, Ministry of Justice, University of Aarhus and University of Copenhagen. The overall aim of DCAW is to contribute to the improvement of animal welfare in Denmark. DCAW contributes to improving animal welfare by presenting an overview of animal welfare-related data from authorities, the farming industry and research, which enables stakeholders to get an overview of the level of animal welfare in various farm animal products. Hence, a key goal is communicating animal welfare knowlegde to relevant stakeholdes, such as farmers, politicians, veterinarians, researchers and the general public. Secondly, DCWA initiates and supports animal welfare research. Since 2010, 13 research projects have been initiated focusing on welfare in farm animals. In 2011, 'methods to measure animal welfare' and 'animal welfare in an economic context' have been the main topics for the research projects initiated. Results from the initiated projects are each year presented at an annual conference hosted by DCAW. At the conference new developments and findings relevant to animal welfare issues are also presented. Finally, a main priority is communicating and collaborating with European and non-European countries. DCAW wishes to share information and ideas about the national control of animal welfare as well as new animal welfare initiatives. An international collaboration will provide a common contribution to the improvement of animal welfare.

Tina Birk Jensen, Louise Holm, Lise Tønner, Unit of Animal Welfare, Danish Veterinary and Food Administration, Mørkhøj Bygade 19, Søborg, Denmark

Sir James Dunn Animal Welfare Centre

Sir James Dunn Animal Welfare Centre: animal welfare research, service and education at the Atlantic Veterinary College

The Sir James Dunn Animal Welfare Centre (SJDAWC) at the Atlantic Veterinary College (AVC), University of Prince Edward Island (www.upei.ca/awc), exists to promote animal welfare by generating and disseminating impartial and scientifically based knowledge and understanding of animal welfare issues in all species. The SJDAWC undertakes animal welfare research and education, and provides information and advice to industry, government, organisations, and the public. The SJDAWC seeks funding for animal welfare research projects and service activities; serves as a resource centre to compile and generate information relevant to animal welfare; and strives to raise the awareness of the public and the veterinary profession on broad questions of animal welfare and to provide accurate, scientifically based information on these questions. The SJDAWC seeks partnerships with industry, government, farming, welfare and veterinary organisations, and with foundations to undertake animal welfare research and service activities. The major sponsor of the SJDAWC is The Christofor Foundation. This funding supports administration and outreach activities, including newsletters and conferences. Animal welfare research and service projects are funded following an annual competition. Research funding can provide research costs and a graduate stipend (animalwelfare@upei.ca). Potential graduate students can also apply for AVC scholarships. The Chair in Animal Welfare and the Coordinator of the SJDAWC undertake undergraduate and graduate teaching in animal welfare and participate in national and regional committees on animal welfare issues.

Michael S. Cockram[1], Alice Crook[2], [1]Chair in Animal Welfare, [2]Coordinator – Sir James Dunn Animal Welfare Centre, Atlantic Veterinary College, University of Prince Edward Island, 550 University Avenue, Charlottetown, PEI, C1A 4P3, Canada

Tshwane University of Technology

Animal ethics at Tshwane University of Technology

The Tshwane University of Technology (TUT) in South Africa was established in January 2004 with the merger of three smaller tertiary institutions based in and around Pretoria. The unique focus of TUT as a university of technology is based on the development of career-focused skills and competencies, at the diploma, four year Bachelors degree and postgraduate levels. TUT supports research and innovation in carefully selected focus areas. The policy of the centralised TUT Research Ethics Committee (REC) states that its mandate is to independently evaluate, approve and monitor research that involves humans, animals and the environment within a framework of generally accepted research ethics guidelines. REC is a standing sub-committee of the Central Research and Innovation Committee. Where research proposals involve the use of vertebrate animals (laboratory, farm and/or wild animals), the proposals are referred to the Animal Research Ethics Committee (AREC) for evaluation and approval. It primarily reviews proposals for animal-related research conducted in the Faculty of Science, which includes the Natural Sciences, Health Sciences and Agricultural Sciences. The AREC is registered with the National Health Research Ethics Council in South Africa and predominantly follows the animal ethics guidelines of the Medical Research Council (SA) and the South African National Standard (SANS 10386 of 2008). Seven permanent members serve on the AREC, including a veterinarian, laboratory animal technologist, farm animal researcher, wildlife expert, SPCA animal welfare officer, and an independent community representative.

Soné Steenkamp-Jonker[1], Francois Siebrits[2], [1]Department of Biomedical Sciences, Faculty of Science, Tshwane University of Technology, Pretoria, South Africa, [2]Chair: Animal Research Ethics Committee, Tshwane University of Technology, Pretoria, South Africa

Scientific abstracts

Session 1. Farm animal welfare assessment and sustainability: connecting social and environmental aspects of measuring welfare

Session 2. Lessons learned from new and existing welfare assessment schemes

Session 3. Development and validation of animal welfare assessment or audit protocols

Session 4. Our experience beyond the farm: animal welfare assessment for zoos, laboratories and animal shelters

Session 5. New and emerging methods for animal welfare assessment

Session 6. Working with end-users of animal welfare schemes

Session 7. Poster session

Farm animal assessment and sustainability

Mitloehner, F., University of California, Davis, Department of Animal Science, One Shields Ave., Davis, CA 95616, USA; fmmitloehner@ucdavis.edu

By examining the historical trends in livestock production in the US, it becomes clear that there has been a marked improvement in efficiency, leading to reductions in numbers of animals required to produce a given amount product that satisfies the nutritional demands of society. For example, researchers at Cornell University found that compared to 1944, the 2007 U.S. dairy industry reduced its greenhouse gas emissions per unit of milk by 63%. Today, one California cow produces the same amount of milk per year as five of her Mexican peers! This reduction was achieved through improved nutrition, management, genetics, etc. born through scientific research that has lead to dramatic improvements in milk production per cow. According to the UN FAO, this type of intensification of livestock production provides large opportunities for climate change mitigation and can reduce deforestation to establish pastures, thus becoming a long-term solution to more sustainable livestock production. Indeed, the UN FAO is currently working on a paper titled 'Shrinking the Shadow', which will focus on how advanced biotechnologies, improved genetics, nutrition, and comprehensive waste management already utilized in most parts of the developed world can be applied effectively worldwide. While the extraordinary reduction in the US dairy industry's carbon footprint may be viewed by some as a vindication of modern production practices, attention should be given to the areas of opportunity that still exist, including transition cow management, lameness, and reproductive failure. Improving these and other areas on US dairy farms should allow for further reductions in carbon footprint per unit of milk, and these areas often intersect with another hot issue that livestock industries face: animal welfare.

Session 1; Tuesday August 9th 8:55 - 12:15 Theatre 2

(To be announced)

Use of farmer focus groups to explore compliance issues relating to a welfare scheme for suckler beef cattle

Dwane, A.M.[1], More, S.J.[1], Blake, M.[2] and Hanlon, A.J.[1], [1]School of Agriculture, Food Science and Veterinary Medicine, University College Dublin, Belfield, Dublin 4, Ireland, [2]Department of Agriculture, Fisheries and Food, Agriculture House, Kildare Street, Dublin 2, Ireland; andrea.dwane@ucd.ie

On-farm animal welfare is primarily determined by farmer compliance with regulations (and quality assurance standards). The levels of compliance however may prove difficult to assess. Ireland's Animal Welfare, Recording and Breeding Scheme for Suckler Herds ('Suckler Scheme') was launched by the Department of Agriculture, Fisheries and Food (DAFF) in 2008, providing financial incentives to suckler beef farmers for implementing a number of specified welfare practices. Initial uptake of the scheme was widespread with approximately 50,000 farmers joining (approximately 76% of eligible herds). Using the Scheme as a case study, our research has focused on exploring the factors that influence beef farmers' willingness or ability to comply with welfare guidelines. In this study, four focus groups (each comprising 7-9 farmers) were conducted in 2009. Participants were sourced through local veterinarians and invited to attend. Audio recordings of discussions were transcribed verbatim and then 'coded' for topics and views mentioned. Coding and thematic analysis were carried out using NVIVO 8. When asked to list times when welfare is most at risk, participants responded as follows: calving (88%), weaning (69%), nutrition/ body condition during pregnancy (25%), handling (25%), housing (22%) and disbudding/ dehorning (16%). All four groups commented that all existing scheme measures were relevant to good farming practices and therefore should continue. Participants suggested amendments to a number of measures as follows: change disbudding ages (59%); simplify paperwork (47%); change the training (38%); increase scheme payments (31%); and change weaning rules (28%). Focus groups provide scope for deeper exploration into the attitudes and beliefs underlying participants' answers in a way that more conventional surveys may not. They also provide an opportunity for farmers to discuss compliance issues without the risk of incurring a farm inspection or financial penalty. Farmers seem motivated to comply when criteria result in financial benefit, are practical and workable, impact positively on welfare and health (e.g. minimum age at first calving and meal-feeding), or are such that failure to comply may trigger inspection and/ or loss of payments (e.g. attendance at training). Farmers seem less motivated to comply when criteria seem impractical at farm level, may have a negligible or even negative effect on welfare (e.g. ages for disbudding), are over-complicated (e.g. paperwork), lack consistency, or may cause financial loss (e.g. weaning).

Expert opinion on animal welfare and non-compliance with legislation in Swedish animal husbandry

Hultgren, J.[1], Algers, B.[1], Blokhuis, H.[2], Gunnarsson, S.[1] and Keeling, L.[2], [1]Swedish University of Agricultural Sciences, Department of Animal Environment and Health, P.O. Box 234, SE-532 23, Sweden, [2]Swedish University of Agricultural Sciences, Department of Animal Environment and Health, P.O. Box 7068, SE-750 07 Uppsala, Sweden; jan.hultgren@slu.se

This study aims to assist the Swedish competent authority in the development of a scheme for risk classification of animal operations under official animal welfare (AW) control. The study evaluated different views on AW and their importance for perceived AW risks. To cover different perspectives of dealing with AW, three groups of experts were nominated to be contacted and recruited in a systematic stepwise process; stakeholder (S), authority (A) and academic (U) experts. A total of 28 S experts from 23 stakeholder organizations or companies, 15 A experts from 6 national or regional control authorities, and 14 U experts from 4 universities or institutes were selected. In a 60-point questionnaire, the experts provided information on themselves, their background, personal views on AW and its interpretation in official control. The experts also assessed AW risks in 177 different categories of Swedish animal husbandry operations. For each category, experts were asked to individually provide values related to a) the probability of non-compliance with current AW legislation found at a fictitious control visit to a randomly selected operation, motivating corrective action (P1), and b) the probability of one or several severe AW deficiencies at a randomly selected operation during a period of one year (P2). This contribution is based on data from the first 40 experts (20 S, 11 A and 9 U) sampled at the first three workshops. P1 and P2 were strongly correlated (Spearman's rho=0.76, P<0.0001). Linear mixed models of P1 and P2 (arcsine transformed) were used to assess associations with expert group (S, A, U) and with questionnaire data, specifying expert id as a random effect. From 66 potential predictors, a handful of variables were retained in each model, including main husbandry type (120 categories; P<0.0001). Of the total variation in P1 and P2, 47-51% resided on the level of the expert, but there was no significant difference between S, A and U experts. Back-transformed least-squares means of probabilities for not complying with legislation or having a severe welfare deficiency (P1:P2) for main husbandry types were highest for mink fur (25:37%) and egg producers (31:27%), intermediate for beef (22:23%), pork (21:19%) and broiler chicken (17:21%) producers as well as for pig (22:23%) and poultry (21:24%) abattoirs, and lowest for pet dog owners (14:14%) and animal clinics/hospitals (6.7:9.3%). We conclude that it is possible to use expert opinion to get a balanced view of AW risks in animal husbandry.

AssureWel: incorporation of welfare outcomes into UK laying hen assurance schemes

Main, D.C.J. and Mullan, S., University of Bristol, Clinical Veterinary Science, Langford House, Langford BS40 5DU, United Kingdom; d.c.j.main@bristol.ac.uk

The potential value of using formal welfare outcomes within farm assurance (FA) schemes has been widely recognised. However, practical challenges have limited their widespread implementation. AssureWel, a collaborative project led by University of Bristol, RSPCA and Soil Association, aims to overcome these challenges in several livestock species. The project will use outcomes to supplement existing standards and inspection processes rather than to produce an overall farm level evaluation such as Welfare Quality®. The new procedures should help schemes deliver better assurance and crucially promote genuine welfare improvement. Focusing initially on laying hens, the project has developed inspection procedures that are feasible and scientifically robust. The first step has been to define a set of 'core' measures that will be assessed on all farms at every inspection (feather loss, cleanliness, beak trimming, measures of aggression and sick/injured birds). This selection will be regularly reviewed to consider including 'aspirational' parameters, such as Qualitative Behavioural Assessment. The principle has been to standardise, where possible, with the relevant parameters in the Welfare Quality ® Protocols. The sampling strategy has also been defined by estimating the 95% confidence intervals (CI) in farm, and overall FA scheme, for sample sizes between 20 and 2000 birds per farm. The trade-off between the time constraints and the usefulness of the information resulted in agreement of a sample size of 50 birds per farm. This gives useful information about the whole FA scheme (maximum 95% CI of 1.8%) but less precise information about individual farms (maximum 95% CI of 28.3%). Repeatability amongst farm assurance assessors has also been addressed during the development stage. Six FA assessors were trained in relevant measures and their inter-observer reliability was tested by assessing live birds on-farm two months later, prior to the start of their farm visits. The recorded prevalences by assessors ranged between 28% and 42% (n=40 birds) for feather loss, 10% and 30% (n=20 birds) for abnormal beak shape and between 40% and 70% (n=20 birds) for comb/head marks. In future, training of all assessors aim to use feedback from repeated reliability tests to ensure adequate standardisation. Welfare assessment alone is unlikely to deliver welfare improvement without some additional driver to motivate farmers to respond positively to the assessment. Hence a strategy has been developed that promotes an interest in the results during the visit and encourages farmers to take action after the visit. If significant problems are observed the scheme's certification process would require improvement on the farm.

Lying behaviour as a welfare indicator on North American freestall farms

Barrientos, A.K.[1], Von Keyserlingk, M.A.G.[1], Galo, E.[2] and Weary, D.M.[1], [1]University of British Columbia, Animal Welfare Program, 2357 Main Mall, Vancouver, BC V6T 1Z4, Canada, [2]Novus International Inc., 20 Research Park Drive, 63304, USA; ale_k_barrientos@hotmail.com

The amount of time cows spend lying down has been considered a welfare indicator on dairy farms, in part because farms with low lying times are believed to be at higher risk for high rates of lameness. We assessed lying behaviour in one group of high producing multiparous Holstein cows for 122 herds in California, the north-eastern United States (New York, Pennsylvania and Vermont) and British Columbia, Canada. Data were collected by the same 2 trained individuals on every farm. Electronic data loggers were attached to 40 randomly selected high production multiparous cows in each herd and used to record lying time and number of lying bouts for 3 d. Hock injuries were evaluated for these same cows on a scale of 1 to 3 (1=healthy and 3=evident swelling or severe lesion). All the cows in the group were gait scored using a 5-point numerical rating system where 1 and 2 are considered non-lame, ≥3 clinically lame, and ≥4 severely lame. Percentage of lame and severely lame cows on farms was not related to lying time, number of lying bouts, or average bout duration, but was positively associated with variation in lying time (SD lying time; $P<0.01$ for both % lame and % severe lame) and the SD of the number of lying bouts ($P<0.05$ and $P<0.01$ for %lame and %severe lame, respectively). Percentage of cows with hock injuries (hock scores 2 or 3) was higher on farms with lower lying times ($P<0.001$) and higher variation in lying times ($P<0.01$). Measures of lying behaviour were strongly associated ($P<0.005$) with the presence of deep-bedded stalls (e.g. 10.4 vs. 11.0±0.1 h/d lying time; 10.8 vs. 9.7±0.2 lying bouts/d; 64.8 vs. 74.3±1.4 min bout duration on farms without versus with deep-bedded stalls). In summary, variation in lying behaviour was associated with the risk of lameness and hock injuries on dairy farms.

Measuring up to expectations: the use of animal outcome indicators in welfare assessment

Keeling, L.J., Swedish University of Agricultural Sciences, Department of Animal Environment and Health, P.O. Box 7068, SE-750 07 Uppsala, Sweden, Sweden; Linda.Keeling@slu.se

There are many reasons why one would want to assess animal welfare. They include wanting information on which to base management decisions, for benchmarking or certification to become a member of an accredited scheme, and to check whether a unit fulfills welfare legislation requirements. As a result, what people may be required to assess will vary. An overall assessment of a farm or unit, for example, will require a combination of animal-based outcome measures covering all dimension of animal welfare, whereas a visit to enforce legislation will need only focus on what is specified in that legislation. Welfare is a characteristic of the individual animal and this is the ground stone for the animal-based approach to welfare assessment. But as feasibility, in addition to validity and reliability of measures, becomes increasingly important, recording animal-based measures is expanding to include new techniques. Valid measures can be increasingly obtained automatically using a pressure pad that the animal has walked over, video image analysis and so on. Furthermore, if the ultimate assessment is to be at the farm or unit level, then herd or flock level measures can be appropriate. Last but not least, if the aim is to assess whether or not there is improvement in animal welfare, tracking change over time becomes important. Benchmarking is already commonly used, but to follow trends will require recording of perhaps a few standardized core animal outcome indicators gathered in a systematic way over several years. In addition to the above considerations, the methodology to assess welfare should reflect whether the intention is to assess welfare at the moment of inspection, or to assess the risk of poor welfare at some point in the future either on that unit or elsewhere. Which methodology will in part be influenced by how often welfare can feasibly be assessed i.e. whether it is an ongoing practice performed by the unit manager, carried out frequently as part of quality assurance, or checked infrequently as is often the practice when enforcing legislation. The less frequent the check the more important that the methodology also addresses risk of poor welfare in the future. The future welfare of an animal is a consequence of the interaction between its current welfare and the environment in which it will be kept until that future point in time. For this reason, predicting future welfare can not only be based on resource and management-based measures but also needs to include relevant animal-based measures. In summary, continued improved knowledge of welfare outcome indicators as well as continued development to improve feasibility are necessary to promote good animal welfare and reduce the risk of poor welfare.

Challenges to implementing animal welfare standards in New Zealand

O'connor, C., Ministry of Agriculture and Forestry, Animal Welfare, P.O. Box 2526, Wellington 6140, New Zealand; cheryl.oconnor@maf.govt.nz

The New Zealand Animal Welfare Act 1999 imposes a duty of care on all owners and persons in charge, to provide for the physical, health and behavioural needs of the animals in their care. The Act provides for the development of codes of welfare and gives legal status to the minimum standards that they contain. Codes are used to promote appropriate behaviour, establish minimum standards and encourage best practice by those in charge of animals. One of the main challenges in developing codes of welfare is to integrate the various, and often conflicting, social, ethical, economic and production management value judgments, with the available science, in a way that does not stifle innovation or require frequent alteration of the codes. This is achieved using outcome-based minimum standards which are designed as animal orientated statements of desired welfare outcomes, accompanied by one or more indicators by which achievement of the outcome can be measured or objectively assessed. The challenge is to develop standards that achieve welfare outcomes consistent with current scientific thinking, are consistent with good practice, meet societal expectations for the welfare of animals, are readily understood and accepted by those who must abide by them, and are effective tools for those who ensure compliance with them. An advantage of outcome-based welfare standards is the freedom they afford persons in charge to develop responses to meet the standard, rather than having an operational standard imposed. One of the key issues in the development of such standards, and assurance programmes, is how welfare can be assessed in an objective, practical and cost-effective way. Codes also have an additional function of providing an educational and informative base that is intended to raise standards of animal welfare. This is achieved by including 'best practice' recommendationswhich are intended to raise the bar above the provision of the animal's minimum requirements alone and aim to drive towards continuous improvement of animal husbandry systems and welfare outcomes. Codes are primarily directed at educating the owners of animals or persons in charge and encouraging their voluntary compliance with their legal obligations and to support industries in the development of compliance and quality assurance programmes. The challenge is to develop a consistent, whole-of-system approach to animal welfare compliance that focuses on interventions which encourage voluntary compliance or deter non-compliance before offending becomes serious. New Zealand's policy is to develop outcome-based welfare standards, to promote and demonstrate maximum voluntary industry compliance with them, and to ensure that any serious breaches are detected and effectively responded to.

Characteristics of dairy herds in different Welfare Quality® categories

De Vries, M.[1], *Van Schaik, G.*[2], *Bokkers, E.A.M.*[1], *Dijkstra, T.*[2] *and De Boer, I.J.M.*[1], [1]*Wageningen University, Animal Production Systems, POB 338, 6700AH Wageningen, Netherlands,* [2]*GD Animal Health Service, POB 9, 7400AA Deventer, Netherlands; marion.devries@wur.nl*

The Welfare Quality® Protocol for Cattle is becoming a standard for assessing dairy cattle welfare in the European Union. It aggregates scores of welfare indicators into a herd classification, assigning herds to an unacceptable, acceptable, enhanced, or excellent category. Our aim was to investigate differences in welfare scores of herds assigned to different categories. Seven trained observers quantified indicators of the Welfare Quality® protocol in 194 Dutch dairy herds (herds sizes from 10 to 211 cows). Herds were non-randomly selected to aim for herds with a lower combined health score based on mortality, udder health, and milk production. Differences in welfare indicator scores among categories were analyzed with Kruskal-Wallis and Chi-square tests. In case of significant differences, categories were compared pairwise. Twenty-five herds were classified as unacceptable, 90 as acceptable, 79 as enhanced, and none as excellent. Herd size, milk production, and type of housing did not differ among categories. Herds classified as enhanced had a maximum health score more often than herds classified as acceptable (32.9 vs. 14.4%, P=0.011). With regard to welfare indicators, herds classified as unacceptable were provided insufficient water more often (P<0.01) and showed higher frequencies of head butts/cow/h (1.17) than herds classified as acceptable and enhanced (0.83 and 0.81, P<0.01). Herds classified as unacceptable showed higher percentages of very lean cows than herds classified as enhanced (7.5 vs. 3.4%, P=0.003), but less cases of mastitis (9.8 vs. 12.9%, P=0.010) than herds classified as acceptable. Compared to enhanced, herds classified as acceptable were provided insufficient water more often (P<0.001), showed higher percentages of cows lying down with collision (39.1 vs. 27.2%, P=0.016), lying outside the supposed lying area (3.0 vs. 1.2%, P=0.003), with nasal discharge (5.3 vs. 2.8%, P=0.009), with mastitis (12.9 vs. 10.9%, P=0.009), with lesions (46.4 vs. 36.6%, P=0.003), could not be approached nearer than 100 cm (25.7 vs. 20.8%, P=0.019) and were less relaxed, content, happy, and more frustrated (P<0.05). Differences in welfare scores between herds classified as unacceptable and acceptable were mainly based on water provision and agonistic behaviors, whereas differences between herds classified as acceptable and enhanced lay in the domain of water provision, use of lying area, health, human-animal relationship and emotional state. Classification of farms might lead to a focus on improving these aspects, and deprive attention from improving scores of other welfare indicators such as lameness.

Welfare risk factor analysis for commercial broiler chickens on farm

Bassler, A.[1], Arnould, C.[2], Butterworth, A.[3], Colin, L.[2], De Jong, I.C.[4], Ferrante, V.[5], Ferrari, P.[6], Haslam, S.[3], Wemelsfelder, F.[7] and Blokhuis, H.[1], [1]Swedish University of Agricultural Sciences, Dept. of Animal Environment and Health, Uppsala, Sweden, [2]INRA, UMR85 PRC; CNRS, UMR6175, Université François Rabelais de Tours, Nouzilly, France, [3]University of Bristol, Clinical Veterinary Science, Langford, United Kingdom, [4]Wageningen UR, Livestock Research, Lelystad, Netherlands, [5]University of Milan, Dept. of Animal Sciences, Milan, Italy, [6]CRPA, Research Centre for Animal Production, Reggio Emilia, Italy, [7]Scottish Agricultural College, Sustainable Livestock Systems, Penicuik, United Kingdom; Arnd.Bassler@slu.se

Aim of the study was to identify risk factors for poor welfare in commercial broiler flocks. Flocks of 91 traditional-intensive broiler chicken farms, located in France, England, The Netherlands, Italy and Denmark, were inspected between 2007 and 2009 using the Welfare Quality® broiler welfare assessment. Resource based measures (RBMs) and animal based measures of welfare (ABMs) were collected by means of a farmer questionnaire and an on-farm inspection protocol. The ABMs were: Mortality, Cleanliness, Gait score (GS), Foot pad dermatitis (FPD), Hock burn, Eye pathologies, Diarrhoea, Qualitative behavior assessment, Avoidance distance test, Touch test and Novel object test. A set of 13-18 RBMs with plausible biological association to each ABM was selected out of a total of 26. Thereafter, two multiple linear regression models were calculated per ABM. Model I: Stepwise selection method (fwd in, bckwd out). Model II: All independent variables that contributed with $P \leq 0.05$ to model I were simultaneously included. 'Country' was tested as potential confounder in all models. Results for model II showed that R^2 ranged from 18 to 63% (mean: 35%). Among the RBMs with individual P-values <0.0005 were (1) flock age, (2) dark period, (3) litter quality and (4) type of heating system: (1) With increasing flock age, the number of birds with a GS >2 ('lame') would rise with 1.1% per day (95% CI: 0.5-1.6). Age range was: 25-49 days, mean 39 days. (2) In flocks with a dark period ≥4 h/d at 3 weeks of age, the number of birds with a GS >2 at the day of inspection would be 11.9% (95% CI: 5.8-18.1) lower than in flocks with a dark period <4 h/d. (3) With decreasing litter quality (scored 1=dry/flaky to 5=wet/capped), the number of birds with moderate or severe FPD would rise with 12.9% (95% CI: 7.1-18.6) per litter-scoring unit. (4) Houses with localized heating systems would have 14.7% (95% CI: 8.0-21.4) less birds with diarrhoea than houses with ambient heating systems. This explorative study indicates that flock age, length of dark period at 21 days of age, litter quality and type of heating system may be relevant factors for lameness, FPD and diarrhoea, respectively.

Analysis of discomfort of animals

Neijenhuis, F. and Leenstra, F., Wageningen UR, Livestock Research, P.O. Box 65, 8200 AB Lelystad, Netherlands; francesca.neijenhuis@wur.nl

For evaluation of animal welfare policy the level of welfare of animals has to be characterized. On the European level, assessment systems for farm animal welfare have recently been developed, Welfare Quality®, but the protocols require farm visits by experts are time consuming and not yet implemented for policy evaluation. Animal welfare is the result of positive (fulfilment of needs, 'natural behaviour') and negative experiences (surgical interventions, chronic stress etc.). In the context of this paper we use the word discomfort when the animal fails to get positive experiences, or gets negative experiences. 'Discomfort' refers to those forms of the physical and mental health of animals being affected, whose nature and existence can be established and substantiated scientifically. We designed a method to characterize discomfort of an animal category by expert view in a semi quantitative way to evaluate trends in the level of discomfort across years on the national level. The method looks at discomfort as experienced by the animal according to animal scientists, with discomfort being evaluated from a scientific perspective. Based on the WQ® system causes of discomfort are indicated and the level of discomfort is scored according to severity and duration for the individual and proportion of the population experiencing this discomfort. We used three categories: 0, 1 or 2 for severity, duration, and the share of the population, describing strictly which value to award when. Per species 2-5 experts came to consensus on scores. Multiplication of the 3 figures identifies items causing severe discomfort of longer duration for a large part of the population. We used the method for several species. Per species between 10 and 40 items were identified. For example, in the 2007 analysis for cattle, pigs, poultry, mink and horses: High discomfort scores for dairy cows include the house design, e.g. restricted cubicles, slippery floors, mobility problems. The main issues for horses are the negative consequences of lack of knowledge among horse owners, social isolation, individual housing, lack of exercise and the mismatch between the breeding purpose (inherent predisposition) and the use of most horses. Low-stimulus surroundings and restricted space cause severe discomfort for pigs, veal calves, mink and poultry. Although confronted at first, the field recognised the results. When repeated, for each item differences in severity, duration and/or (most likely) proportion of the population can be identified. The overall picture can provide an indication for a trend in level of discomfort for the species. Assessment of discomfort of animals on a national level by an expert view can be used for evaluation of animal welfare policy and for prioritization of policy.

Evaluation of the sample needed to accurately estimate outcome-based measurements of dairy welfare on farm

Endres, M.I.[1], Espejo, L.A.[1] and Tucker, C.B.[2], [1]University of Minnesota, 1364 Eckles Avenue, St. Paul, MN 55108, USA, [2]University of California, 1 Shields Avenue, Davis, CA 95616, USA; miendres@umn.edu

Dairy welfare assessment programs are becoming more common on U.S. farms. Outcome-based measurements such as locomotion, hock lesion, hygiene, and body condition scores are included in these assessments. The objectives of this study were to describe the prevalence of parameters, and to investigate the proportion of cows on a farm needed to provide an accurate estimate of locomotion, body condition, hygiene, and hock lesion scores. We visited 52 randomly selected freestall dairy farms in Minnesota once during the summer. Cows in the high-production pen were evaluated for lameness using a 5-point locomotion scoring system (1 = normal locomotion, 2 = imperfect locomotion, 3 = lame, 4 = moderately lame, and 5 = severely lame). Animals were scored for body condition (BCS) using a 5-point scale, where 1 = emaciated and 5 = obese. Hygiene scores were assessed by the amount of dirt on the udder and lower hind legs based on a 5-point scale with 1 = clean and 5 = dirty. Hock injuries were classified as 1 = no lesion, 2 = hair loss (mild lesion), and 3 = swollen hock (severe lesion). Cows were rarely very thin (BCS = 2; 0.10% of cows) or very fat (BCS ≥4, 0.11% of cows). A relatively large portion of the cows were dirty (hygiene score 3 or more = 54.9% of cows). Approximately a quarter (24.4%) of the cows were classified as lame (locomotion score ≥3) and 6% were classified as severely lame (score ≥4). Ten percent of cows had severe hock lesions (10.6% cows with hock score = 3). Subsets of data from each farm were created with 10 replicates of random sampling with replacement using PROC SURVEY SELECT in SAS. These subsets represented 100, 90, 80, 70, 60, 50, 40, 30, 20, 15, 10, 5 and 3% of the cows measured/pen. The association between the estimates derived from each subset and the estimated prevalence of the pens was evaluated using linear regression. Recording 15% of the pen represented the percentage of clinically lame cows (score 3 or more) with high accuracy ($R^2>0.9$), although a higher percentage (30%) of the pen needed to be measured in order to accurately estimate severe lameness (score 4 or more). Only 15% of the pen needed to be sampled to accurately estimate ($R^2>0.9$) the percentage of the herd with hygiene score ≥3, whereas 30% needed to be scored to accurately estimate the prevalence of severe hock lesions. Estimating the portion of thin and fat cows required that 70 to 80% of the pen be measured in order to accurately describe this parameter. Thus, unsurprisingly, a higher percentage of the group must be sampled to generate accurate estimates for relatively rare parameters among lactating cattle (e.g. very thin cows).

Can monitoring water consumption at pen level detect changes in health and welfare in small groups of pigs?

Seddon, Y.M., Farrow, M., Guy, J.H. and Edwards, S.A., Newcastle University, University of Newcastle, NE1 7RU Newcastle Upon Tyne, United Kingdom; yolande.seddon2@ncl.ac.uk

The monitoring of water consumption is a tool that might be useful to detect changes in health and welfare and monitor management. This has often been done through recording at the building level. This study explored whether monitoring the daily consumption of water at the pen level was able to indicate changes in the health and welfare within the group. The water consumption of 24 pens of 10 mixed gender finishing pigs was monitored from entry to a fully-slatted finisher building (52.6±0.5 kg) until slaughter (85.3±0.6 kg). Water was available ad libitum via two nipple drinkers per pen. A water meter was fitted to the supply to each pen, and logged the flow rate over 5 minute intervals. Sensors also logged internal and external temperatures. Meters linked to a modem (Barn Report, Farmex Ltd., UK), from which the data could be accessed via internet connection. Two types of commercial drinker nipple were fitted equally across pens. Per pen, clinical disease symptoms and treatments administered were recorded daily, to produce weekly pen health scores, and pigs were weighed every two weeks. A diary of daily human activity within the building monitored any disturbance. A multiple linear regression model determined factors responsible in part for the variation in the total daily water consumption of pigs at a pen level. The initial model included: internal and external building temperature, the number of pigs in the pen, the drinker type, the liveweight of the pen. A Generalized Linear Model was used to determine whether the water intake per pig per day from each week on trial was related to current and subsequent cases of clinical disease as defined by health score. The number of pigs and their weight were added as covariates. The drinker type, the number of pigs per pen, mean pig liveweight and environmental temperature were significant factors affecting total variation in daily pen water consumption ($P<0.001$ for all) and accounted, in combination, for 43% of the total variation (R^2=43.4%). Disturbance to pigs caused by human activity did not explain any further variation in daily water consumption and neither did current episodes of clinical disease. However, the quantity of water consumed per pig per day in the second week after building entry differed in relation to the number of pigs with clinical disease seen in the following week (0 pigs: 5.4, 1 pig: 5.0, 2 pigs: 4.3±0.25 l/day/pig, $P<0.05$). Differences in daily water intake can be linked to subsequent changes in health status of small groups of pigs, suggesting this might be a sensitive measure of subclinical disease.

Qualitative behaviour assessment in sheep: consistency across time and association with health indicators

Phythian, C.J.[1], Wemelsfelder, F.[2], Michalopoulou, E.[1] and Duncan, J.S.[1], [1]University of Liverpool, Leahurst, CH64 7TE, United Kingdom, [2]Scottish Agricultural College, Bush Estate, EH26 0PH, United Kingdom; elemicha@liverpool.ac.uk

Qualitative Behaviour Assessment (QBA) is a 'whole-animal' methodology that addresses welfare by scoring an animal's body language using terms such as content, anxious or relaxed. This method has shown to be a reliable and feasible indicator for on-farm welfare in pigs, cattle, poultry and sheep, however the consistency of on-farm QBA over time has so far not been addressed. The aim of this study was to investigate whether and how on-farm QBA of sheep varies over the different seasons of the year, and/or is associated with other sheep health indicators. A trained assessor visited each of 12 farms six times within one year at 2 month intervals, and at each farm assessed either the entire flock, or, if farms were large, a group of ± 100 sheep selected ad hoc (assuming homogeneity within the flock). QBA was based on a list of 12 descriptors previously developed for sheep in collaboration with Quality Meat Scotland and the Scottish Society for the Prevention of Cruelty to Animals. After completion of QBA, the number of sheep with signs of lameness and breech dirtiness was counted. QBA scores from the 6 assessment dates were analysed together using Principal Component Analysis (PCA – correlation matrix, no rotation). The effect of assessment-date on PCA farm scores was analysed with repeated-measures analysis of variance. The association of PCA farm scores with the proportion of lame and 'dirty rear' sheep was examined by Spearman's rank correlation coefficient. PCA distinguished two dimensions of sheep expression: PC1 (48.2% variation) ranging from 'content/relaxed/thriving' to 'distressed/dull/dejected' (summarized as 'mood'), and PC2 (19.81%) which ranged from 'anxious/agitated/responsive' to 'relaxed' (summarized as 'responsiveness'). No effect was found of assessment-date on PC1 ($P<0.31$), indicating the sheep's general mood to be relatively stable throughout the year. There was an effect of assessment-date on PC2 ($P<0.001$), indicating a possible 'relaxing' effect of the presence of young lambs on the sheep's responsiveness. Proportion of lame sheep in a group was negatively correlated with farm scores on PC1 ($n=72$; $rho=-0.59$, $P<0.001$), indicating an association of negative mood with lameness. No other significant correlations were found. Farm scores on PC1 ('mood') are used as a QBA welfare indicator in Welfare Quality® and Scottish farm assurance protocols. By demonstrating the relative stability of such sheep farm scores over time, and their meaningful association with lameness, these results further support the reliability and validity of QBA as an indicator for on-farm welfare assessment.

Developing measures for auditing welfare of cattle and pigs at slaughter

Grandin, T., Colorado State University, Animal Sciences, 350 West Pitkin Street, Fort Collins CO 80523-1171, USA; cheryl.miller@colostate.edu

Since 1999, animal welfare auditing programs that utilize five numerically scored core criteria have been used successfully by major restaurant chains to monitor animal welfare in slaughter plants. The use of objective numerical scoring resulted in great improvements because the slaughter plant managers knew exactly what was required to pass a welfare audit in order to remain on the approved supplier list. They had to achieve specific numerical scores. The five scored criteria (critical control points) are: (1) Percentage of animals stunned effectively with one application of the stunner, (2) Percentage rendered insensible when hoisted to the bleed rail (has to be 100% to pass the audit), (3) Percentage of cattle or pigs vocalizing in the stunning box or restrainer, (4) Percentage of animals that fall down during handling, and (5) Percentage of animals moved with an electric prod. Audit data collected in 2010 by two restaurant companies in 30 beef plants, indicated that 77% of them effectively stunned 100% to 99% of the cattle with a single shot from a captive bolt gun. All 30 plants passed the audit, which required 95% of more of the cattle stunned with one shot. Before the program started in 1999, baseline data showed that only 30% of the beef plants were able to effectively stun 95% of the cattle with a single shot. The main cause of poor stunning was lack of stunner maintenance. In 2010, the percentage of cattle vocalizing in the stunning area of all the plants was 5% or less. This is a huge improvement compared to baseline data, which showed that the worst plant had 32% of the cattle vocalizing due to excessive pressure from a restraint device. In 95% of the beef plants, and 88% of 25 pork plants, 0% of the animals fell during unloading movement in the lairage and during handling in the stunning area. The five numerically scored items are critical control points because they measure multiple problems and they are indicators of major welfare problems. For example, a high percentage of animals vocalizing could be due to electric prod use, excessive pressure from a restraint device, or slamming a gate on an animal. The welfare audit also contains a number of secondary items that are on the checklist, but a passing score is required on all five of the core criteria to pass the welfare audit. The author has served as an instructor to train auditors and inspectors. Most of these workshops for training auditors last 1 ½ days and include two plant visits. In order to train people effectively during this short workshop, the audit has to be simple. All scores are per animal, and involve a categorical (yes/no) score.

Monitoring farm animal welfare during mass euthanasia for disease eradication purposes

Berg, C., Swedish University of Agricultural Sciences, Dept of Animal Environment and Health & Swedish Centre for Animal Welfare, POB 234, 532 23 Skara, Sweden; Lotta.Berg@slu.se

The most efficient way of preventing animal welfare problems during disease eradication operations is to ensure high levels of biosecurity, well-designed surveillance programmes and rapid alert systems for detecting possible outbreaks of different contagious diseases. This way, the need for mass euthanasia will be minimized, and thereby the risk of poor animal welfare during such operations. Nevertheless, disease outbreaks are likely to occur at irregular intervals and may affect several farms in an area or sometimes entire regions or countries. Mass euthanasia will primarily affect infected herds/flocks of farm animals, but may also be used for contact herds where it is considered too risky to wait for analysis results or for healthy herds which have to be euthanized for animal welfare reasons, for example when routine transport and slaughter cannot be carried out. The main focus of virtually all contingency plans during a disease eradication operation, especially for zoonotic diseases, will be on preventing further spread of the disease in question to humans or other animals. Speed is essential, great financial value is at stake and the situation can be very traumatic for both authority staff and farmers involved. Animal welfare hasn't always been considered a priority in these situations. However, there isn't necessarily a contradiction between depopulation efficiency and animal welfare considerations. The ethics of killing large numbers of animals can certainly be debated, but this presentation will rather focus on how to avoid unnecessary animal suffering once the authorities have made the decision to apply depopulation as the main strategy during a disease outbreak. Animal welfare considerations must be incorporated in contingency planning and emergency preparedness, organization, training, reporting and follow-up. How can such operations be carried out without unduly compromising animal welfare, how should this be monitored and how can we ensure that lessons learned are applied in future handling of such events? In Europe, upcoming legislation (EC 1099/2009) requests actions plans to ensure compliance with animal welfare regulations before commencement of the operation, to safeguard the welfare of the animals in the best available conditions. It will require that reports on depopulation operations are made publicly available via the Internet, including information on the reasons for the depopulation, the number and species of the animals killed, the methods used, description of the difficulties encountered and, where appropriate, solutions found to alleviate or minimize the suffering of the animals concerned etc. Developing such monitoring systems is an urgent task.

Use of video recordings, a vehicle tri-axial accelerometer and GPS system to study the effects of driving events on the stability and resting behaviour of cattle, young calves and pigs

Spence, J.Y.[1] and Cockram, M.S.[2], [1]Humane Slaughter Association, The Old School, Brewhouse Hill, Wheathampstead, Herts, AL4 8AN, United Kingdom, [2]University of Prince Edward Island, Sir James Dunn Animal Welfare Centre, Department of Health Management, Atlantic Veterinary College, 550 University Avenue, Charlottetown, PEI, C1A 4P3, Canada; mcockram@upei.ca

The welfare of animals in transit (e.g. risk of injury) may be affected by driving events such as acceleration, braking and cornering and the manner of their performance. The relationships between driving events and the behavioural responses of the animals to these events were examined. A single-deck, non-articulated vehicle was fitted with a video recording system and a vehicle monitoring system (GPS and tri-axial accelerometer). Two different drivers [D] each drove three standard journeys (two 3-h stages [S] on different types of roads [R]) for each animal type. Six different groups of: five cattle (370 kg live weight, space allowance 1.13 m^2/animal), ten calves (59 kg live weight, space allowance 0.34 m^2/calf) and ten pigs (111 kg live weight, space allowance of 0.45 m^2/pig) were each transported on separate journeys. The posture and loses of stability of two randomly selected focal animals from each journey were observed. Mixed model and generalized linear models were used to study the effects of D, R and S. The percentages of losses of balance and falls preceded within 5 s by each type of driving event were calculated. Cattle stood still for >97% of each part of the journey. Young calves spent more time lying down during the second S (12%) than during the first S (5%). There was a significant D×R×S interaction ($P<0.01$) on the time that pigs spent lying down (0 to 30%). However, they spent more time sitting down and this time was greatest on motorways (43%) and during the second S (37%). The fewest losses of balance occurred during the motorway stages (D×R×S interactions $P<0.001$ for cattle and young calves and R×S interaction $P<0.05$ for pigs). There were about 2 falls for cattle and pigs, and 5 falls for calves per journey. Longitudinal acceleration of up to 0.1 g preceded 71, 33 and 64% of falls in cattle, calves and pigs, respectively. Braking of up to 0.1 g preceded 14, 23 and 36% of falls in cattle, calves and pigs, respectively. Lateral acceleration of >0.1 to 0.5 g preceded 57, 50 and 57% of falls in cattle, calves and pigs, respectively. Travelling on motorways will encourage livestock to rest and reduce any discomfort from repetitive driving events. Anticipating potential driving events and preparing for them will reduce the likelihood and severity of losses of stability. A driver training DVD was produced from this study.

Assessing welfare during transport: relationships between truck temperatures, pig behaviour, blood stress markers and meat quality

Brown, J.[1], *Crowe, T.*[2], *Torrey, S.*[3], *Bergeron, R.*[4], *Widowski, T.*[4], *Correa, J.*[5], *Faucitano, L.*[6] *and Gonyou, H.*[1], [1]*Prairie Swine Centre, Saskatoon, S7H 5N9, SK, Canada,* [2]*University of Saskatchewan, Dept. of Chemical & Biological Engineering, 105 Administration Place, S7N 5A2, SK, Canada,* [3]*Agriculture and Agri-Food Canada, 50 Stone Rd E, N1G 2W1, ON, Canada,* [4]*University of Guelph, Animal and Poultry Science, University of Guelph, Guelph, N1G 2W1, ON, Canada,* [5]*Canadian Meat Council, 955 green Valley Cr., Ottawa, K2C 3V4, ON, Canada,* [6]*Agriculture and Agri-Food Canada, 2000 Rue College, Sherbrooke, J1M 1Z3, QC, Canada; jennifer.brown@usask.ca*

The effects of transport conditions on pig behaviour, welfare and meat quality were assessed under Canadian conditions. Temperature was monitored in a triaxle potbelly livestock trailer during 11 trips transporting market pigs to a large commercial packing plant. A total of 2,145 pigs were shipped, with 5 loads in winter and 6 in summer. Each trip took approximately 7 h. To monitor trailer temperatures, logging devices were mounted from the ceiling, with 5 devices per compartment in 9 compartments. Pigs' postures during transport were assessed using photographs taken at 5 minute intervals by digital cameras mounted in 7 compartments. Blood samples were collected from a subsample of 330 pigs (30 per load) at slaughter, and meat quality was evaluated in the longissimus dorsi (LD) and semimembranosus (SM) muscles of 470 pigs. Behaviour data from half of the winter trials was lost due to technical difficulties. Correlation analysis (Pearson) showed that in hot temperatures more pigs lie down (r= 0.450; P<0.001) and fewer pigs stand (r= -0.473; P<0.0001). The motivation behind postural adjustments is unclear; pigs may lie in warmer temperatures due to fatigue, and/or may prefer to stand at cold temperatures to minimize heat loss. A mixed model analysis in SAS showed significant effects of season and compartment on posture. More pigs laid down during transport in summer than in winter (43 ±2% vs. 30 ±3%, respectively; P<0.01) and more pigas laid down in rear top compartments (53±4%) and fewer in top front compartments (27±3%; P<0.001). The postural differences were not explained by temperature differences between compartments, and may instead result from variation in vibration and truck forces. Higher truck temperatures were associated with lower levels of blood lactate and CPK at slaughter, a higher number of fighting-type marks on the carcass, lower pH at 24 h post-mortem (pHu) in the SM muscle and lower pHu and higher drip loss in the LD muscle (P<0.0001). In conclusion, a better understanding of transport conditions and their effects on pigs and pork quality merits further work on both welfare and economic grounds. The increased availability of new technologies makes this work feasible.

Animal welfare and different pre slaughter procedures in Uruguay

Del Campo, M.[1], Manteca, X.[2], Soares De Lima, J.[1], Brito, G.[1], Hernández, P.[3], Sañudo, C.[4] and Montossi, F.[1], [1]INIA, R5 Km386, 45000 Tbo, Uruguay, [2]Univ. Aut. Barcelona, Bellat., 01893, Spain, [3]Univ. Polit. Valencia, C de Vera s/n, 46022, Spain, [4]Univ. Zaragoza, Servet 177, 50013, Spain; mdelcampo@tb.inia.org.uy

From an ethic perspective and as a meat exporter country, Uruguay must consider public sensitivity to animal welfare (AW). Scientific data are lacking on stress-inducing factors in cattle extensive systems, transport and lairage, so the aim of this study was to evaluate the effect of diet, temperament, transport and lairage time, on AW. Sixty Hereford (H) and Braford (B) steers were finished in 2 diets: D1) native pasture plus corn grain (H n=15, B n=15) and D2) high quality pasture (H n=15, B n=15). Live weight (ADG) was registered each 14 days and also Temperament by 3 tests: Crush score, Flight time and Exit velocitiy, building a multicriterial Index (Tindex). Cortisol, creatine kinase (CPK) and non sterified fatty acids (NEFA) in blood were used to assess stress at the farm, after transport, after lairage and pre slaughter. Steers were slaughtered in two groups (50% of animals from D1 and 50% from D2 in each group) after 15 and 3 hours in lairage pens. The journey lasted 3.5 hours (commercial truck and proper driving). Behaviour was evaluated during lairage by direct observation (scan sampling). Fighting and mounting (F&M) were also registered between 2 scan periods (behaviour sampling). Carcass pH was registered at 24 hs post mortem (pm) in the Longissimus thoracis muscle. Mixed models adjusted for repeated measures were used to study major effects on liveweight, temperament and blood indicators through time, and non parametric tests were used for behaviour analysis. ADG and temperament did not differ between diets ($P<0.05$). B steers were more excitable showing lower values of Tindex than H steers (62.10±4.10 in H vs. 50.90±4.00 in B). Calmer animals had higher ADG within both diets and breeds and lower values of stress indicators at all pre slaughter stages ($P<0.05$). Any health problem occurred during the experiment. Each step suggested higher stress once leaving the farm, but transportation did not affect the adrenocorticotropic axis activity. No differences were found in F&M frequency ($P<0.05$) during the first hour in pens, but animals that remained 15 hours became calmer afterwards, reaching lower and adequate final pH values (5.68±0.04 vs.5.83±0.04 in the short lairage group). In conclusion, negative effects of transport may be reduced after short travels by proper handling/equipment; temperament appears to be a valid tool to improve productivity and to reduce the physiological stress response at all pre slaughter stages; less than 3 hours in lairage should affect AW and also meat quality if animals are stressed.

Body space requirements of broilers depend on density and compression allowed

Koene, P.[1], De Boer, I.J.M.[2] and Bokkers, E.A.M.[2], [1]Wageningen UR Livestock Research, Animal Sciences, Marijkeweg 40, 6709 PG Wageningen, Netherlands, [2]Wageningen University, Animal Production Systems, POB 338, 6700 AH Wageningen, Netherlands; paul.koene@wur.nl

There is continuing debate about the space needs and requirements of broiler chickens. The aims of this study were to measure the amount of floor area a 6-week-old broiler occupies for different behaviours and to use the obtained results in two models to estimate the number of birds that can be kept per m^2 in large flocks. The first model computed the space needed per bird performing a behaviour in relation to a flock size of 20,000 broilers; the second model computed stocking density based on synchronisation of 7 behavioural elements and the actual body space. Pictures (frames) were saved of overhead video recordings of broilers (2.468 kg on average) kept in floor pens of 1 m^2 with either 8 (low density) or 16 birds (high density) per pen. Individual body space (cm^2) was measured in pixels from these pictures for seven behaviours. Density affected body space of the behaviours idle ($F = 18.65$, $P=0.005$), drinking ($F=9.34$, $P=0.022$), ground pecking ($F=34.20$, $P=0.001$) and preening ($F=22.30$, $P=0.003$). In high density less space was available for these behaviours. Density did not affect stretching ($F=3.24$, $P=0.122$) and walking ($F=0.18$, $P=0.684$). For dustbathing differences tended to be significant ($F=53.28$, $P= 0.087$), but the number of records was low. This was probably because dustbathing occupied the most space (630-762 cm^2) and could not be well performed, so only few samples were found in the video recordings of the low ($N=7$) and especially the high ($N=2$) densities. Based on body space projection only and leaving out movements and social space requirements both models showed a limit to stocking density of broilers. The first model, computing space needed per bird performing a behaviour in relation to flock size, showed that 15.3-15.7 birds/m^2 (37.8-38.7 kg/m^2) can be housed maximally based on low density measurements and 18.5-19.4 birds/m^2 (45.7-47.9 kg/m^2) based on high density measurements. The second model, computing stocking density based on synchronisation of behaviour and body space, showed that 13.7-15.9 birds/m^2 (33.8-39.2 kg/m^2) can be housed maximally based on low density measurements and 15.4-18.6 birds/m^2 (38.0-45.9 kg/m^2) based on high density measurements. Results based on high density measurements implied that birds are compressed. Given the restrictions of a limited number of behaviours and no inclusion of movement and social interactions in the models of this study, stocking density in large flocks should not exceed 16 birds/m^2 (39.4 kg) because that would lead to compression of birds which will suppress opportunities for behavioural expression and therefore impair welfare.

Kinematic assessment of broiler leg health

Caplen, G., Hothersall, B., Colborne, G.R., Nicol, C.J., Murrell, J.C. and Waterman-Pearson, A.E., University of Bristol, School of Veterinary Sciences, Langford House, BS40 5DU, United Kingdom; gina.caplen@bristol.ac.uk

Broiler gait is currently assessed via a broad qualitative scale. As part of a study investigating lameness-associated pain in broilers we used kinematic analysis as a quantitative means of studying gait. A comparison between lame (Gait Score (GS)3; n=10) and non-lame (GS0; n=10) Ross broilers was made and the methodology then used to assess NSAID treatment (30 mg/kg oral Carprofen dosed 2 h prior to testing) on lame birds (GS3; n=10). 3-D back and leg positional data were collected using reflective markers and a ProReflex® infrared camera system over several runs. Gait parameters were measured in paired (left then right leg) steps (SL: stride length; SD: stride duration; %ST: percentage stance; VL: vertical leg; VB: vertical back; TB: transverse back displacement). Values for each paired stride were (1) averaged to produce a mean value (e.g. SL_m) and (2) subtracted from one another to calculate 'difference' (e.g. SL_d). To control for velocity, and account for repeated measures, data were analysed using multi-level models generated with MLwiN 2.22 software. Three levels were utilised, with each stride nested within run, within bird. For each parameter the likelihood ratio test (LRT) was used to compare the fit of two specific models, one with and one without the test factor ('GS' for the first study and 'treatment' for the second). In the first study gait score was a significant negative predictor of SL_m (LRT=25.836, P<0.001), SD_m (LRT=24.852, P<0.001), and VL_m (LRT=3.838, P=0.05), and a positive predictor of TB_m (LRT=4.70, P=0.03). Lame birds took shorter, quicker steps than non-lame birds and lifted their legs less, but had more transverse back movement. Gait score was also a significant negative predictor of VB_d (LRT=10.388, P=0.001), and positive predictor of TB_d (LRT=5.647, P=0.017): lame birds had less variation in vertical (greater variation in transverse) back movement between paired steps than non-lame birds. In the second study carprofen treatment was a significant positive predictor of both SD_m (LRT=4.185, P=0.041) and $\%ST_m$ (LRT=3.835, P=0.05), indicating that post-treatment, birds had longer stride duration and spent a greater percentage of each stride in the stance phase. Carprofen treatment was also a significant negative predictor of $\%ST_d$ (LRT=4.14, P=0.042), indicating that lame birds demonstrated less variation in stance between paired steps following treatment. Correlation of quantitative gait differences with visual assessment of leg health is encouraging. It is hoped that this technique will prove an important tool for monitoring subtle improvements in leg health following analgesic treatment (various types/doses are being tested), or breeding programmes designed to reduce lameness.

A protocol for measuring foot pad lesions in commercial broiler houses

De Jong, I.C., Van Harn, J., Gunnink, H., Van Riel, J.W. and Lourens, S., Wageningen UR Livestock Research, P.O. Box 65, 8200 AB Lelystad, Netherlands; ingrid.dejong@wur.nl

Foot pad lesions are preferably measured at the slaughter plant. However, foot pad lesions will be included as welfare indicator in the Broiler Directive in The Netherlands and therefore lesions in flocks that will be slaughtered abroad need to be assessed at the farm itself. The aim of the present experiment was to develop a protocol for measuring foot pad lesions in broiler houses. Points to investigate were the inaccuracies of measurement with different number of locations in the house, and different number of birds sampled per location. In addition, it was tested if samples taken near the walls differed from samples taken at other locations in the house, and if there were differences between the first five birds sampled at a location, and all birds sampled at that particular location. Samples were taken in eight randomly selected commercial flocks (Ross 308) with varying severity of foot pad lesions. Samples were either taken at 4, 8 or 10 locations (randomly selected), with 5, 10, 20 or 25 birds per location. Foot pad lesions were scored in all birds for both feet using the Swedish scoring method, i.e. score 0 for no lesions, score 1 for mild lesions, and score 2 for severe lesions. There were no significant differences in severity of foot pad lesions between the first five birds and all birds sampled at a particular location. There were also no significant differences in severity of foot pad lesions between locations near the walls or between locations in the central part of the house. However, there was a significant effect of sampling location ($P<0.001$), which means that the severity of foot pad lesions is unevenly distributed over the house. The data were used to construct a model showing the inaccuracy related to the number of locations sampled in the house, and the number of birds sampled per location. The model shows that in situations with at least 4 locations differences in inaccuracy are relatively small when a total of 100 birds or more is sampled. However, the inaccuracy rapidly grows from 4 to 1 sampling locations. Inaccuracy is largest in a flock with variation in foot pad scores, as compared to flocks with little variation (i.e. >90% score 0, or >90% score 2). In conclusion, a) sampling at least 100 birds in a house is preferred, and b) it is preferred to sample as many locations as possible (e.g. 10 locations with 10 birds per location are preferred over 4 locations with 25 birds per location). The results of this experiment can be used to determine the optimal sample size in a commercial broiler house, taking into account an acceptable level of inaccuracy, the work load for the assessor and disturbance of the birds.

Conflicting expectations to welfare inspections expressed by Danish farmers

Anneberg, I.[1], Sandøe, P.[2], Sørensen, J.T.[1] and Vaarst, M.[1], [1]Department of Animal Health and Biosciences, Aarhus University, Blichers allé 20, 8830 Tjele, Denmark, [2]Faculty of Life Sciences, University of Copenhagen, Bülowsvej 17, 1870 Frederiksberg C, Denmark; inger.anneberg@agrsci.dk

This paper presents results from a study of how Danish farmers perceive welfare inspections carried out by Danish authorities. Such inspections are carried out for regulatory purposes on at least 5% of all Danish livestock farms per year. During the unannounced visit, the inspector checks all the provisions of the animal welfare legislation, most of which relate to the physical environment of the animals. In cases where the inspector finds that the animal welfare does not comply with the legislation the results can range from warnings followed by repeated inspection, reductions in subsidies from the EU, fines or being reported to the police. 22 unannounced inspections on farms of different size were observed. The observations were followed by on-farm qualitative interviews with 12 farmers and 12 inspectors, chosen from the inspections on selected large pigs- and dairy cattle farms. The interviews followed a semi-structured interview guide. They were tape recorded, transcribed, coded and analysed using phenomenological analysis. The attitudes of the farmers to the inspections were highly ambiguous: On the one hand the farmers perceived animal welfare inspections as inevitable and necessary, based on their belief that there are some farmers who do not comply with the law. At the same time, the farmers in general felt that the inspections were unfair. The farmers describe the lack of fairness through three main themes: (a) There are too many regulations related to animal welfare today, making it difficult for the farmers to overview what they have to do. Therefore the authorities will almost always be able to find small violations of the law, if they want to. (b) There is too little room for discussion and interpretation during the inspections (even though they at the same time want inspections to be fair, in the sense that all farms are being inspected according to the same standard.) (c) It is unfair when suffering of a few animals counts as a broken regulation on a farm where the majority of the animals are doing fine. The farmers clearly have an ambiguous attitude to welfare inspection: They want the inspections system to be far more transparent, fair and objective, and at the same time they want the outcome of the inspection to be open for human interpretation and negotiation. This ambiguity should be taken into account both by researchers, farmers, their organizations and the authorities, when discussing the use of on-farm welfare assessment for regulatory purposes.

Welfare assessments on zoo animals: past, present, and future

Wielebnowski, N., Chicago Zoological Society, Center for the Science of Animal Welfare, 3300 Golf Road, Brookfield, Illinois 60513, USA; nadja.wielebnowski@czs.org

Over the past decade zoos and aquariums have increasingly become interested in systematic and quantitative welfare assessments of their collections. World zoo organizations, such as WAZA and AZA, have placed the provision of best possible welfare for all animals in their care as one of their highest priorities. Accordingly many managers of zoo and aquarium species are trying to identify measures and mechanisms to ensure that existing and emerging welfare concerns are being identified and addressed as quickly as possible, while continuously striving to increase existing animal care standards and improve upon current knowledge in animal care and welfare. I will discuss the history of zoo animal welfare assessment as well as several recent developments in welfare assessment and welfare research in zoos and aquariums with a focus on AZA (Association of Zoos and Aquariums) accredited zoos. Resource-based and animal-based assessment techniques and tools will be highlighted. Especially some recent developments involving keeper ratings of individual animal welfare and the validation and application of such ratings will be presented. Finally, a perspective on future research needs, directions and developments in zoo animal welfare will be provided.

Research animal welfare assessments in the United States

Bayne, K., AAALAC International, 5283 Corporate Dr., Suite 203, Frederick, MD 21703, USA; kbayne@aaalac.org

In the United States, federal oversight of animal care and use for research, testing, and teaching is achieved by numerous laws, regulations, policies, and guidelines from the following government agencies: the US Department of Agriculture (USDA), the US Public Health Service (PHS), the Department of Defense, the US Food and Drug Administration, the US Environmental Protection Agency, and the Veterans Administration. There are also state and local regulations that apply to protecting the welfare of research animals. Thus, the system of oversight of research animal welfare is matrix-based. The USDA is the lead federal agency for providing oversight of research animal welfare. Compliance with the USDA's Animal Welfare Act Regulations is ensured through unannounced annual inspections, conducted by a veterinarian, of institutions using animals in research. USDA inspection reports are available to the public on the USDA's website. Research funding agencies often require adherence to additional standards. For example, institutions receiving a grant from the National Institutes of Health (or other PHS agency) must conform with federal regulations and the PHS Policy on Humane Care and Use of Laboratory Animals, and are subject to cause and not-for-cause site visits by PHS officials from the Office of Laboratory Animal Welfare (OLAW). Information from OLAW (e.g. correspondence with institutions, annual reports, etc.) is available through the Freedom of Information Act. Issues of noncompliance identified during inspections by the USDA or OLAW can result in a variety of punitive measures including fines, and/or revoking the institution's privilege of using animals in research. Research institutions may also choose to voluntarily participate in an accreditation program implemented by the Association for Assessment and Accreditation of Laboratory Animal Care (AAALAC International). AAALAC International uses standards derived from the peer-reviewed literature to assess the care and use of research animals. AAALAC conducts triennial assessments of animal research programs and confers an accreditation status based on the outcome of those assessments. At the local level, oversight of research animal welfare is accomplished, in part, by the Institutional Animal Care and Use Committee (IACUC). A significant role of the IACUC is the performance of semiannual animal care and use program reviews and inspections of areas where animal are housed and studied. These various audit strategies, ranging from the institutional level to federal oversight, are complementary in their different approaches to promoting research animal welfare. Details of these various audit programs will be discussed, to include a summary of the ways in which research animal welfare is enhanced through these audits.

Development and validation of an audit system to assess welfare on large felids in captivity

De La Fuente, M.F.[1], Gimpel, J.[2], Bustos, C.[3] and Zapata, B.[1], [1]Universidad Mayor, Veterinary School, Camino La Piramide, 32, Chile, [2]Pontificia Universidad Catolica de Chile, Av. Libertador Bernardo O´Higgins #340, 32, Chile, [3]Universidad Santo Tomás, Av. Ejercito #146, 32, Chile; beatrizzapata@hotmail.com

Large felids have serious difficulties to meet their psychological needs in captivity, however the majority of zoos around the world keep at least one specimen of lion. In this study we develop and validate a system to evaluate animal welfare for large felids in captivity. First we identify valid and reliable welfare indicators related to the following principles: Good feeding, Appropriate Housing, Good Health and Appropriate Behavior. In order to select valid and relevant indicators of those principles, updated literature on ecology, behavior, health and housing conditions for Panthera leo, Panthera tigris, Panthera onca and Puma concolor was examined, and resources and animal based indicators were selected. Then a protocol was built, in which each indicator was scored using a three levels scale from 0 to 2, being 0 associated to the poorest welfare and 2 associated to the best welfare. The reliability of the protocol was assessed calculating Kendall coefficient of concordance (W) among four blind trained observers. The protocol was applied in five zoos in central Chile and 30 enclosures. The protocol was validated through questionnaire to veterinarian and technicians in charge of the care of big wild felids. An index of welfare was built based on the optimal score in relation to the observed score and four levels of welfare were defined: Good (0.76-1), Moderate (0.51-0.75), Poor (0.26-0.50) and Very Poor (0-0.25). Thirty four welfare indicators were identified. The Kendall coefficient of concordance was high ($W>0.8$, $P<0.05$), therefore the reliability of the global protocol was adequate for all the enclosures studied from the five zoos. The questionnaire was answered by 15 experts from Chile and also foreign countries (Mexico and Costa Rica). More than 80% of experts found that welfare indicators selected were pertinent and approximately 75% found that the score was well assigned. The assessment of the zoos according our index was variable, finding Low values (0.38) and also Good values (0.81). The developed system to assess welfare of big felids at zoo level showed to be feasible, reliable and valid.

Assessing fear in farmed mink: validity and sensitivity of the 'glove test'

Meagher, R., Duncan, I. and Mason, G., University of Guelph, Animal & Poultry Science, 50 Stone Road E., Bldg. 70, Guelph, Ontario, N1G 2W1, Canada; rmeagher@uoguelph.ca

Fear is a major cause of poor welfare in farm animals, making it crucial to develop valid and reliable methods of screening for fear in large numbers of subjects. For farmed mink in Europe (Scandinavia), the most common method is a 'stick test' where a wooden spatula is inserted into the cage and the minks' immediate responses noted. However, on Ontario farms, fearfulness in this test was very rare, while aggressive responses were prevalent: problematic since the welfare significance of aggression is unclear. We therefore developed and validated a modified test, replacing the stick with a handling glove, the finger of which is inserted into each cage. This test was expected to be more sensitive to fear because such gloves would be associated with past handling, and also smell of minks' stress odour. When both tests were applied to 120 mink on a commercial Ontario farm, the 'glove test' did prove more sensitive for detecting fearfulness than the stick test (McNemar's χ^2=19.2, P<0.0001), as well as successfully reducing aggressive responding (McNemar's test, χ^2=18.2, P<0.0001). Furthermore, in other studies, the test had construct (convergent) validity. Thus in a five minute test conducted on 46 animals, mink immediately classified as 'fearful' then spent more time exhibiting other fear-related behaviours ('ambivalence': $F_{1,9}$=11.06, P=0.009), while mink immediately classified as 'curious' then spent more time in investigation ($F_{1,43}$=23.34, P<0.0001). In a population of 473 mink screened during pregnancy on three commercial farms, in one colour-type, fearfulness in the glove test predicted which females would be barren (Fisher's exact P=0.006). Finally, in an environmental enrichment study, it predicted endocrine responses to improved housing. Faecal corticosteroid metabolite (FCM) output was assessed in 27 mink housed for eight months in either standard or large enriched cages (the latter containing swimming water and 'toys' but also allowing mink to retreat to areas inaccessible by humans). Fearfulness did not predict baseline FCM before the housing change, but fearful mink showed larger decreases in FCM when given enrichment than non-fearful mink did (GLM: housing condition*fear $F_{1,22}$=3.28, P=0.08; fearful vs. not-fearful mink moved to enriched housing: $F_{1,22}$=6.17, P=0.02), suggesting that they benefitted most from these larger, more complex cages. Together, these findings show that modifying a stick test into a glove test yields a practically useful, valid way to assess fearfulness in mink, and one more sensitive than the conventional stick test for detecting low levels of fear.

The development of an on-farm welfare assessment protocol for foxes – WelFur

Mononen, J.[1], Ahola, L.[1], Hovland, A.L.[2] and Koistinen, T.[1], [1]University of Eastern Finland, Department of Biosciences, P.O. Box 1627, 70211 Kuopio, Finland, [2]Norwegian University of Life Sciences, Department of Animal and Agricultural Sciences, P.O. Box 5003, NO-1432 Aas, Norway; jaakko.mononen@uef.fi

In 2009 the European Fur Breeders' Association (EFBA) made an initiative to launch the development of on-farm welfare assessment protocols for farmed mink (Neovison vison), and blue (Vulpes lagopus) and silver foxes (Vulpes vulpes) and their hybrids. These WelFur protocols are based on Welfare Quality® (WQ) principles and criteria. Here we describe the state of affairs for the WelFur fox protocol after 1.5 years development. Reviews of each of the 12 welfare criteria were written to identify the welfare measures that have been used for farmed foxes. The reviews formed the basis for the suggestions of potential measures to be included in the WelFur fox protocol. The suggestions were discussed and modified in four meetings by a group consisting of fur animal scientists, external animal welfare experts (including WQ experts), and EFBA representatives. All potential measures were evaluated for their validity, reliability and feasibility. Altogether 214 measures were identified, but a majority of the measures used in scientific studies were regarded unfeasible. At the moment the suggested fox protocol includes 14 animal-based, 4 resource-based and 5 management-based measures. There is at least one criterion with at least one animal-based measure for each of the four WQ principles, and altogether 7 out of the 12 criteria include animal-based measures. Practically on all fox farms, all three production periods (P1: from pelting to mating, P2: from mating to weaning, and P3: form weaning to pelting) take place on the one and same farm, and the relevance of the various animal-based measures may vary between these periods. This necessitates careful planning of the calculation for the accuracy of the overall scores, and puts challenges on implementation of WelFur, since probably three visits are needed before one gets a final assessment result for a farm. However, it seems that P3 may be the best time for most animal-based measures included in the current WelFur fox protocol. The relevance of some measures may also differ between the two fox species or their hybrids. In addition, the minor differences in fox farming practices between Finland and Norway (the two EFBA countries with fox farming) has to be taken into account. Using WQ project and protocols as a model has been an extremely fruitful approach in developing WelFur protocols and after developing the scoring system in early 2011, the implementation of the whole fox protocol will be tested already in 2011-12. However, at the same time it is apparent there is a need to develop new measures and to refine the old measures.

Farm animal welfare: assessing risks attributable to the prenatal environment

Rutherford, K.M.D.[1], Donald, R.D.[1], Arnott, G.[1], Rooke, J.A.[1], Dixon, L.[1], Mettam, J.[2], Turnbull, J.[2] and Lawrence, A.B.[1], [1]SAC, Animal Behaviour & Welfare, Easter Bush, Midlothian, EH25 9RG, United Kingdom, [2]University of Stirling, Institute of Aquaculture, Stirling, FK9 4LA, United Kingdom; kenny.rutherford@sac.ac.uk

Over the last decade research into the impact that prenatal conditions have on farm animal welfare has expanded rapidly. Across all major farmed species, including poultry and fish, a number of experimental studies have clearly shown that early life experiences have a substantial impact on outcomes of great relevance to later health, welfare and productivity. In particular, stress or sub-optimal nutrition experienced by the mother during pregnancy has been shown to have wide-ranging and important effects on how her offspring develop with implications for how they will cope with their social, physical and infectious environment. Variation in the conditions for development provided by the reproductive tract or egg, for instance by altered nutritional supply or hormonal milieu, may therefore explain a large degree of variation in many welfare and productivity relevant traits. We have reviewed the literature in cattle, sheep, pigs, poultry and fish to identify possible management practices for pre-birth/hatch individuals that could compromise their later welfare. This work also has relevance for the welfare of other animals under human care, in laboratories, zoos or as companion or sporting animals. Overall the findings highlight the importance of extending the focus on animal welfare to include the prenatal period, an aspect which until recently has largely been neglected. This work will benefit the pregnant dam and her developing offspring. Furthermore welfare inspections of, and standards for, breeding units or hatcheries should incorporate detailed information on prenatal hazards to ensure the best welfare outcomes for all individuals affected. At present, such considerations are typically overlooked since offspring outcomes are manifest in the future and often in a different location. There are also implications for how research work in farm animal welfare assessment is conducted. On-farm surveys of animal welfare often find substantial variation within particular farm production systems. Whilst epidemiological analyses of such data often allow many causal explanatory factors to be uncovered, these studies rarely investigate the role of the prenatal environment in determining welfare outcomes. In conclusion, the possible importance of welfare standards for gravid animals, not merely for their own welfare, but for the welfare of the next generation they 'house' is widely unappreciated within most farming systems. Animal welfare could be improved in many farming systems by paying closer attention to how breeding females are housed and managed.

Epidemiological relationships between cage design, feather cover and feather cleanliness of commercial laying hens

Newberry, R.C.[1], Kiess, A.S.[2], Hester, P.Y.[3], Mench, J.A.[4] and Garner, J.P.[3], [1]Washington State University, Box 646520, Pullman, WA 99164-6520, USA, [2]Mississippi State University, Box 9665, Mississippi State, MS 39762, USA, [3]Purdue University, 125 S Russell St, West Lafayette, IN 47907-2042, USA, [4]University of California Davis, One Shields Avenue, Davis, CA 95616, USA; rnewberry@wsu.edu

The influence of critical aspects of cage design on hen welfare outcomes has not been fully assessed in commercial flocks of laying hens under North American conditions. We used an epidemiological approach to identify key features of commercial cage housing that affect feather cover and cleanliness. We obtained data on the feather condition of White Leghorn hens in 167 commercial houses distributed across all regions of the United States using a five-point scale for feather cover and a four-point scale for feather cleanliness. Hens in four adjacent cages were sampled in each of nine locations within each house. In each set of four cages, we scored the hen closest to the observer in two cages and the hen furthest from the observer in the remaining two cages. Four body areas visible from outside of the cage (head/neck, shoulders/upper back, lower back, tail) were scored separately on each scale and results were averaged for each hen. From each house, we also gathered systematic data on cage dimensions and other aspects of the housing environment. We used general linear modeling to identify a statistical model comprising features of the housing environment that best described the variance in feather cover and cleanliness. To avoid confounding effects, we controlled for the effects of all other variables when testing each individual variable. Among producers, the best-fit model explained 50 and 66% of the variation in feather cover and cleanliness, respectively. Feather cover was significantly greater in houses with (1) frequent waste removal, especially houses with an A-frame cage configuration, (2) cup drinkers or plain nipple drinkers rather than nipple drinkers with drip cups, (3) incandescent rather than fluorescent lights, (4) hens that had cleaner feathers, (5) greater floor space allowance per hen, (6) less feeder space per hen, (7) lower cage floor slope, and (8) lower cage height. Feather cleanliness was greater in (1) houses without evaporative cooling, (2) the Hy-Line W36 strain of White Leghorns, (3) shallower cages, (4) drinkers positioned towards the back rather than the front of the cage, and (5) taller cages. These results indicate that multiple aspects of cage design and management affected feather cover and cleanliness, sometimes in unexpected ways. Our study emphasizes the value of an epidemiological approach for identifying factors to take into account when designing systems that optimize hen welfare under commercial conditions.

Milk recordings and meat inspection data as a diagnostic tool for identifying dairy herds with welfare problems

Otten, N.D.[1], Toft, N.[1], Thomsen, P.T.[2] and Houe, H.[1], [1]University of Copenhagen, Faculty of Life Sciences, Large Animal Science, Groennegaardsvej 8, 1870 Frederiksberg C, Denmark, [2]University of Aarhus, Faculty of Agricultural Sciences, Animal Health and Bioscience, Blichers Alle 20, 8830 Tjele, Denmark; nio@life.ku.dk

Register data such as meat inspection, milk recording and mortality data holds great potentials in dairy science depending on the study purpose. In recent years, there has been an increased demand on cheap welfare indicators which can identify livestock herds with potential welfare problems. The objective of this study was to evaluate a set of indicators from register data as a diagnostic tool for the prediction of animal welfare at herd level using lameness as a gold standard. A total of 39 large dairy herds where thorough clinical examinations including lameness scorings had been performed were selected for the study. A lameness prevalence >13% was used as a gold standard for classifying the herd as having 'poor welfare'. A set of seven production and management related variables from the Danish National Cattle Database were used. These indicators were in previous studies found significantly correlated with lameness prevalence and were easily feasible from the register. Three cut off levels were assigned to each of these indicators all depending on three stages of severity of each indicators' excess above cut off (mild, moderate or alarming excess) concerning their impact on welfare. In order to evaluate the most suitable cut off level and predictive model an overall summation of indicator scores were combined in a set of six different definitions of 'poor welfare' based on the sum and severity of the indicator cut off levels, (e.g. a minimum of two indicators had to be above the lowest cut off level). Sensitivity and specificity were calculated for both the cut off levels and definitions. The summary indicator definition claiming three indicators being above the lowest cut off level showed the best sensitivity at 94%, while the best specificity at 74% was obtained when the definition required four indicators being above the highest cut off level. For optimum Se/Sp (the definition maximizing both Se and Sp) the 'poor welfare' definition was four indicators above the middle cut off level. When individual indicators were ranked in accordance to their predictive qualities, mortality and lean cows at slaughter were the most predictive indicators being able to correctly classify 64% and 56% of herds, respectively. The results indicate that register data could hold a potential in identifying livestock herds at risk of having welfare problems, but further refinement of a combination of indicators is needed before an operational model is ready.

Automated monitoring of animal welfare indicators

Rushen, J.[1], Chapinal, N.[2] and De Passille, A.M.[1], [1]Agriculture and Agri-Food Canada, Pacific Agri-Food Research Centre, 6947 Highway 7, Agassiz, BC V0M 1A0, Canada, [2]Ontario Veterinary College, Population Medicine, Guelph, Guelph, ON, N1G 2W1, Canada; Jeff.Rushen@agr.gc.ca

On-farm scoring of indicators of good or poor welfare, especially behavioural ones, is challenging during an audit of a farm since these occur irregularly over time. The increasing availability of low cost technology now makes automated monitoring of animal behaviour and some aspects of animal physiology feasible. Lameness in farm animals (a common cause of poor welfare) is associated with a number of behavioural changes, which can be monitored automatically. For cows and pigs, these include changes in the way that animals distributes their weight when standing or walking, reductions in number of visits to an automated milking or feeding systems, and reduced activity. Image analysis can also help identify changes in gait that may indicate lameness. These measures have been shown to have a reasonable degree of specificity and sensitivity in identifying lame animals, and can be used to assess the degree of pain associated with lameness, especially when different measures are combined. Automated monitoring of changes in general activity (time spent standing, number of steps taken etc.) and visits to automated feeders can also be used to identify animals that are becoming ill, and this can be confirmed by remote recording of body temperature by infra-red telemetry. Automated monitoring of general activity may also be used to assess the presence of other risks to animal welfare, such as poor design of stalls for cattle, but in these cases the link to animal welfare is sometimes less clear. Less research has been done into automatic monitoring of indicators of positive welfare, or of social behaviour. Technological developments have provided us with a variety of tools that can be used to monitor behaviour automatically, and these have great potential to improve our ability to monitor animal welfare indicators on-farm. However, it is important that these measures be validated, and that animal welfare assessment schemes not place undue emphasis on behavioural indicators solely on the basis that they can be monitored automatically

Non-invasive measures of stress for monitoring animal welfare: possibilities and limitations

Palme, R., Institute of Chemistry and Biochemistry, Department of Biomedical Sciences/University of Veterinary Medicine, Veterinärplatz 1, 1210 Vienna, Austria; rupert.palme@vetmeduni.ac.at

A multitude of endocrine mechanisms is involved in coping with challenges. Glucocorticoids (cortisol and/or corticosterone), secreted by the adrenals, are front-line hormones to overcome stressful situations. Their concentration in plasma samples is widely used as a parameter of adrenocortical activity and thus of disturbance. Unfortunately, blood sample collection itself disturbs an animal. During the last several years non-invasive methods for the measurement of glucocorticoids or their metabolites have become increasingly popular. Pros and cons of various sample materials that can be collected in a non-invasive or minimal invasive way (saliva, excreta, milk and hair) for the determination of glucocorticoids will be discussed. Above all, fecal samples offer the advantage that they can be collected easily, safely and stress-free. In fecal samples circulating hormone levels are integrated over a certain period of time. As a consequence fecal glucocorticoid metabolites represent the cumulative secretion and they are less affected by short episodic fluctuations of hormone secretion. However, in order to gain reliable and valuable information about an animal's adrenocortical activity, certain criteria have to be met: Depending whether the impact of acute or chronic stressors is assessed, frequent sampling might be necessary whereas in other cases, single samples will suffice. Background knowledge regarding the metabolism and excretion of glucocorticoids is essential and a careful validation for each species and sex (sometimes even for young and mature animals) investigated is obligatory. In addition, this presentation will address analytical issues regarding sample storage, extraction procedures, and immunoassays and various examples of a successful application will be given. Applied properly, non-invasive techniques to monitor stress hormone metabolites in fecal samples of various species are a useful tool for welfare assessment, especially as they can be easily applied at farm or group level.

Development of a novel and automated measure of affect in pigs

Statham, P., Campbell, N., Hannuna, S., Paul, E., Colborne, R., Browne, W. and Mendl, M., University of Bristol, Langford, BS40 5DU, United Kingdom; poppy.statham@bristol.ac.uk

Most people's concerns for animal welfare are likely based on the assumption that animals can experience negative subjective affective states and hence suffer. Although we cannot measure such states directly, to get a truly accurate assessment of on-farm welfare, we need to develop validated proxy indicators of animal affect. 'Defence cascade' (DC) responses offer one possibility. They are triggered by unexpected stimuli and involve an initial 'startle, a monitoring phase (freeze), then either resumption of ongoing behaviour or fleeing. Critically, they are modulated by affective state in humans and rodents. Pigs show a clear DC response to unexpected stimuli and here we examine whether pharmacological induction of a calm state can modulate this, and if it can be automatically quantified using computational image analysis (IA). 12 female pigs (mean 85 kg) were housed in pairs. Each pair was exposed to two treatments: i.m. injection of both pigs with (1) an anxiolytic midazolam (M) 0.15 mg/kg and (2) saline control (C). Treatments were separated by 96 h and order was balanced across pairs. At 20, 35 and 50 mins following injection on each test day, the sound of a bursting balloon was administered to the pigs in their home pen and their responses video recorded. There was no effect of treatment on the observed activity of the pigs prior to administration of the sound (e.g. standing/rooting: M (24/36 tests), C (25/36 tests); walking: M (3/36 tests), C (5/36 tests), chi-square on all measures=10.2, df=7, NS), or on IA analyses of pixel movement at this time (M=3.46±0.54, C=4.67±0.67, F1,4=0.799, NS), indicating no baseline differences in activity. However, repeated-measures ANOVA with treatment as a between-subjects factor showed that M pigs exhibited a significantly smaller initial startle reaction, measured by IA analysis of pixel movement during 1 s following the sound stimulus (M=7.08±1.08, C=14.70±1.44, F1,4=53.1, P=0.002), and a significantly shorter freezing response (M=0.37±0.26, C=2.07±0.59, F1,4=9.5, P=0.037) and time to resume ongoing activity (M=0.90±0.37, C=4.86±0.89, F1,4=10.96, P=0.03), both measured in seconds. The application of an anxiolytic did not appear to reduce the activity of the pigs overall, however it significantly decreased the DC response. These findings clearly support the idea that individuals in a less anxious (more positive) state show attenuated startle and freeze components of the DC response. The DC response thus shows promise as a new indicator of affective state and welfare in pigs, and it can be measured by IA techniques. Further studies have been funded to examine this relationship in more detail and develop the IA techniques required to take this measure onto farms.

Applying welfare training in global commercial settings

Butterworth, A.[1] *and Whittington, P.*[2], [1]*University of Bristol, Clinical Veterinary Science, Langford, BS40 5DU, United Kingdom,* [2]*Animal Welfare Training, The Long House,Banwell, BS29 6BW, United Kingdom; andy.butterworth@bris.ac.uk*

Animal 'keepers' are influenced by how animals are viewed in their own society and how much 'permission to care' they feel granted to them. Levels of 'care' can be subsumed by commercial pressures, lack of time, perceived lack of 'value' for individual animals, perception of animal issues, or sometimes, through lack of exposure to concepts of animal care and welfare. Welfare training can be; Interesting; many are truly interested in what they do and respond positively to 'training' which adds to their knowledge and capacities. Compulsory; some businesses demand training in animal care issues as a part of retailer requirements or social responsibility coverage. Challenging; peoples views on animal issues are varied and there is unlikely to be universal agreement or acceptance of animal welfare topics. Training may start at a 'fundamental' level with concepts of pain and stress, stockmanship and care. In other places, this knowledge is 'assumed' and training covers specific elements, e.g. biosecurity, stock management, transport and slaughter. EU Directive 2007/43/EC requires training of poultry keepers in physiology, drinking and feeding needs, animal behaviour, concepts of stress, practical aspects of handling of chickens, catching, transport, emergency care, biosecurity. Training may be 'capacity building' 'activity which strengthens the knowledge, abilities, skills and behaviours of individuals and improves institutional structures and processes such that organisations can efficiently meet their goals in a sustainable way.' The key words are knowledge, abilities, skills, behaviour of individuals, improvement, 'sustainable'. Training is moulded by the needs of the audience so 'one size fits all' is not usually appropriate, but, there may be some rules which can be applied; Do not start by importing values and technology or procedures which those trained won't and can't use. Understand why the people you train do what they do, and why they keep doing it. Initial training must be sympathetic to local knowledge and resources. Engage with the industry and its affiliates, they were here before you, and will remain as the active forces after you have gone. If possible, obtain government, professional and academic support and involvement. Beware – in the absence of knowledge and training, new technologies and procedures can create new welfare problems. Training in animal care and welfare can be challenging and is part of a gradual process of involvement in hearts, minds, attitudes and social norms: be patient, it may take some time.

Tools for improved animal welfare in Swedish dairy production

Hallén Sandgren, C., Winblad Von Walter, L. and Carlsson, J., Swedish Dairy Association, Box 210, 101 24 Stockholm, Sweden; jonas.carlsson@svenskmjolk.se

In 2004 Swedish Dairy Association (SDA) started the development of a Scheme for Animal Welfare. The aim is to increase the welfare of the animals, to strengthen the profitability in Swedish dairy herds and safeguard consumer trust. A research project has resulted in a model to distinguish between herds with high and low chance for good welfare by using a combination of 7 indica-tors from pre-collected register data including two fertility measures, mortality in three age categories and incidence of mastitis and feed-related diseases. Two different animal welfare tools, Animal welfare signals and Ask the cow, are now available. The tools are intended to improve animal welfare in dairy herds, and to enhance the knowledge and understanding of animal welfare among farmers and their employees. The web report Animal welfare signals, which can be reached through the SDA website, was introduced in 2010. The report is available for all dairy farmers affiliated to the milk recording system and has until March 2011 been used by 70% of all dairy farmers. It consists of indicators from register data shown to be associated to on farm welfare assessment and/or production economy. The indicators are presented as traffic light smiley´s and offers possibilities of benchmarking with herds with optional characteristics for herd size, breed, organic, production and milking system, etc. Ask the cow, introduced in 2010, is an advisory animal welfare service performed by a trained technician. The service includes assessment of animal-based welfare measures of cows, young stock and calves. The results from the assessment are compared with that of randomly selected Swedish dairy herds and are presented as flowers. Broken petals indicate animal welfare improvements are needed and the service includes an action plan based on 'stair case principles' which emphasize that fulfilment of basic needs is a prerequisite before aiming at higher levels of welfare. For example, if feed provision is scarce, it is irrelevant to change the feed ration. The tools are extensively used by veterinarians in preventive health pro-grammes and by farmers and extension officers to strengthen the profitability in dairy production. Ask the cow is applied by the dairies to safeguard welfare and consumers trust. The model classifying herds as having high or low chance for good welfare is now tested in cooperation with the authorities. The work and the tools will be presented at the congress.

Trade-off between animal welfare and cost price: broiler catching as a case

Eilers, K.[1], Mourits, M.[2] and Bokkers, E.[1], [1]Wageningen University, Animal Production Systems, POB 338, 6700AH Wageningen, Netherlands, [2]Wageningen University, Business Economics, POB 8130, 6700EW Wageningen, Netherlands; Eddie.Bokkers@wur.nl

Improving animal welfare within livestock production seems inevitably related to extra production costs for which farmers rarely receive compensation. Sustainable improvements are therefore often prevented. By providing insight in the cost price of welfare improving alternatives, it becomes possible to communicate about costs and price implications with farmers, retail, and consumers. We studied animal welfare and costs of alternatives for catching broilers ready for slaughter. Catching broilers is done by catching crews who lift 6-8 broilers upside down at a time by their legs and put them in containers. This method of catching results in injuries e.g. bruises, broken wings and legs, and therefore impairs broiler welfare. In addition, manually catching broilers is a heavy, dirty, and unhealthy job. Four alternatives were elaborated for an average Dutch broiler farm (7 rounds/yr, 120,000 birds/round) and compared with traditional catching as a reference situation (one catching crew of 10 people): (1) alternative manual catching (two birds lifted upright at one time), (2) use of a 2nd catching crew replacing the 1st crew half time, (3) use of a catching machine in ownership, (4) hiring a catching machine. Animal welfare aspects were studied using literature. Cost aspects were analysed using partial budgeting. Alternative manual catching was the most animal friendly method. Machine catching resulted in less or comparable number of injuries compared to traditional catching. Birds dead on arrival (%) was higher for machine catching than for manual catching. All alternatives gave rise to extra costs per bird in comparison with the reference: (1) € 0.090, (2) € 0.001, (3) € 0.005 or € 0.0001 (depending on type of catching machine), (4) € 0.020. Sensitivity analyses showed that for large farms buying a catching machine is profitable. Wages for a catching crew had the largest impact on the alternative manual catching method. Price of a catching machine may be around € 100,000 to compete with manual catching. A high catching speed of the machine reduced costs but impaired welfare; catching speed should be around 8,000 birds/h to compete with manual catching. Introduction of alternative catching methods maintaining or increasing carcass quality did not result in extra returns for an average farm. Larger farms may benefit from catching machines, although the impact on carcass quality is unclear. Although increased costs per bird were small, cumulated over total flock size they cannot be borne by farmers alone. Introduction of an alternative catching method is only possible when extra costs for farmers are paid, e.g. by consumers or retail.

Evaluation of the preference for different free-stall bedding systems by dairy cows under field conditions

Burger, Y. and Jahn-Falk, D., Clinic for cattle, Hofbieber, Germany, Am Kiesberg 14, 36145 Hofbieber, Germany; info@rinderklinik-hofbieber.de

It was the aim of this study to evaluate the preference of dairy cows for different free-stall surfaces under field conditions. The study took place in a commercial free-stall barn, where three different types of free-stall bedding systems were installed: 14 cubicles were filled with a 12 cm deep bed consisting of short-cut straw mixed with calcium-carbonate. In 12 cubicles comfort-mattresses were installed and 28 cubicles presented rubber mats as a surface. Every cow in the herd had access to each of the 54 free-stalls. 2 hours after morning-milking (1st visit) and 2 hours after feeding (2nd visit) the stall use (cows lying) was evaluated. The cow-comfort-quotient, the cud-chewing-index and the stall-standing-index were calculated. The cows significantly preferred the straw-filled stalls. The average (Ø) percentage of cows found lying in these cubicles was 91.88%. The stall use of the mattress-stalls was Ø 58.01% and of the rubber mat-stalls Ø 42.86%. The Ø cow-comfort-quotient (CCQ) of the straw-filled stalls was Ø 95.06%, whereas the CCQ of the mattress-stalls (Ø 80.60%) and the rubber mat-stalls (Ø 79.99%) was significantly lower. Only the CCQ of the straw-filled stalls was higher than the required limit of 85%. The evaluation of the cud-chewing-index resulted in no difference between the three bedding systems. The stall use found during the 1st visit was significantly higher (Ø 76.26%) than during the 2nd visit (Ø 55.13%). But only the stall use of the mattress- and rubber mat-stalls were lower during the 2nd visit. The difference was Ø 78.79% to Ø 42.24% for the mattress-stalls and Ø 78.79% to Ø 32.76% for the rubber mat-stalls. The overall CCQ did not change significantly. Only for the rubber mat-stalls the CCQ was significantly lower in the 2nd visit. During the 2nd visit more cows were found ruminating (Ø 50.80% to Ø 39.41%) except for the rubber mat-stalls. In this study cows showed a significant preference for the free-stalls filled with short-cut straw. Stall use can be an indicator of cow preference and preference can be interpreted as a measure of cow comfort. Cow comfort is an important factor concerning animal health and rate of replacement. Therefore the cow comfort contributes to the economy of dairy farming.

Modelling the welfare of growing-finishing pigs: a meta-analytical approach

Averos, X.[1], Brossard, L.[1], Edwards, S.A.[2], Dourmad, J.Y.[1], De Greef, K.H.[3], Terlouw, C.[4] and Meunier-Salaün, M.C.[1], [1]INRA, UMR SENAH, Domaine de la Prise, 35590, Saint Gilles, France, [2]Newcastle University, School of Agriculture, Food and Rural Development, Agriculture Building, NE1 7RU, Newcastle Upon Tyne, United Kingdom, [3]Wageningen UR, Animal Sciences Group, P.O. Box 65, 8200 AB Lelystad, Netherlands, [4]INRA, UR1213 Herbivores, -, F-63122, Saint-Genès-Champanelle, France; Marie-Christine.Salaun@rennes.inra.fr

In the EU 'QPorkChains' Project, existing scientific knowledge has been integrated into a series of models to predict the effects of different swine production systems on animal welfare, taking into account the animal, farmer, and consumer-citizen perspectives. From the perspective of the pig, complexity was reduced by hypothesizing that any production system may be defined by means of a limited number of factors. A meta-analytical approach was used to determine the effects of these factors on the welfare of growing-finishing pigs with respect to their resting, exploring, and feeding activities. Results indicate that the physical and social environment (space allowance, average temperature, floor characteristics and group size), the presence of bedding, the presence, number, presentation and characteristics of point-source enrichment objects, the feeder characteristics (presence of drinking water, protection and space/pig at the feeder), feeder design (individual/collective), feed characteristics (net energy and protein content) and feeding levels (*ad libitum*/restricted), and the individual characteristics of animals (BW, sex and genetics) may simultaneously influence these different activities. The existence of interactions between these factors was also determined for a number of indicators relative to the behaviour and the performance of growing-finishing pigs, highlighting the multidimensionality of the welfare issue. A key future perspective will be to take into account the variability between the pigs within a group, allowing prediction of effects of the different characteristics of production systems on the behaviour and performance of growing-finishing pigs at the population level, rather than as a simple mean individual value.

Test-retest repeatability of the Welfare Quality® assessment protocol for growing pigs on intensive farms

Temple, D.[1], Dalmau, A.[2], Manteca, X.[1] and Velarde, A.[2], [1]UAB, School of Veterinary Science, 08193 Bellaterra, Spain, [2]IRTA, Finca Camps i Armet s/n, 17121 Monells, Spain; deborah.temple@uab.cat

Test-retest repeatability is defined as the agreement between results on the same assessment conducted at two different times by the same observer. Test-retest repeatability of each animal-based measure included into the Welfare Quality® protocol was evaluated on 15 intensive farms of growing pigs. An average of 11 months elapsed between the two visits and no changes in management routines or housing conditions were undergone by the farmers during this interval. Animal-based measures included under the principles good feeding, good housing and good health were scored at pen or individual level according to a three-point scale ranging from 0 (good welfare) to 2 (poor or unacceptable). Appropriate behaviour was assessed by means of scan sampling of social and exploratory behaviour, qualitative behaviour assessment (QBA) and the human-animal relationship test. Data from the QBA were analysed by principal component analysis (PCA) and expressed at farm level. A Wilcoxon signed rank test was used to test whether the mean results obtained during the two visits were significantly different. For all animal-based measures, partial Spearman correlations were calculated removing the effect of the age of the animals. Moderate bursitis (P=0.015), widespread skin discolouration/inflammation (P=0.002), positive social behaviour (P=0.0004) and the QBA (P<0.01) presented mean results significantly different between the two visits. Test-retest repeatability was relatively high for pig dirtiness on less than 50% of the body (rs=0.72) and for the human animal relationship test (rs=0.69). As well, huddling (rs=0.49) and the first principal component of the PCA from the QBA presented a moderate repeatability between the two visits (rs=0.52). The remaining measures were not repeatable from one visit to the other. These results indicate that the majority of farms surveyed did not present a persistent welfare problem. The incidence of each measure at farm level should be studied within batch and between batches. Whether this variability is controlled or not by the farmer should be assessed. The repeatability of a measure at farm level gives an idea of the chronicity of a welfare problem and should be carefully considered in a welfare assessment.

Thermal imaging to detect stress and empathic responsiveness in chickens

Edgar, J., Paul, E. and Nicol, C., University of Bristol, Clinical Veterinary Science, Langford House, Langford, Bristol, BS40 5DU, United Kingdom; j.edgar@bris.ac.uk

The phenomenon of stress-induced hyperthermia (SIH) has been described in numerous species and is characterised by an increase in core body temperature and a decrease in surface temperature within minutes of the onset of 'emotional stress'. Thermal imaging has traditionally been used as a sensitive means of assessing health status in a variety of domestic species, including horses, cattle and pigs. However, few studies have assessed the value of thermal imaging to detect SIH. The aim of the current study was to investigate whether thermal imaging could detect measurable changes in surface body temperature in chickens during (1) exposure to a mild stressor and (2) their chicks' exposure to a mild stressor. 14 domestic hens were exposed to two replicates of the following conditions in a counterbalanced order: Control (C; hen and chicks undisturbed), Air Puff to Chicks (APC; air puff directed at chicks at 30 second intervals), Air Puff to Hen (APH; air puff directed at hen at 30 second intervals) and Control with Noise (CN; noise of air puff at 30 second intervals). During each test, thermal images of each hen's head were taken at one-minute intervals throughout a 10-minute pre-treatment and a 10-minute treatment period. A repeated measures ANOVA was conducted with Condition and Period as within-subjects factors. Post-hoc tests (LSD) were conducted in the event of a significant interaction effect from an ANOVA, during which pre-treatment and treatment periods were compared for each condition. Hens responded to both APH and APC with a decrease in eye temperature (P=0.010). Comb temperature decreased exclusively in response to APH (P=0.003). Temperature changes during both treatments were associated with an increase in time spent standing alert (P=0.000) and a decrease in preening (P=0.000). No such changes occurred during any control period. It was concluded that thermal imaging of the eye and comb may be viable as a non-invasive, sensitive indicator of stress in chickens. Further research should focus on methods to determine the valence of the response and on the application of thermal imaging within the wider context of poultry welfare research.

The impact of lameness on the level of pain and decrease in productivity of finisher pigs using expert opinions

Birk Jensen, T.[1], *Halkjær Kristensen, H.*[2] *and Toft, N.*[2], [1]*Danish Veterinary and Food Administration, Mørkhøj Bygade 19, 2860 Søborg, Denmark,* [2]*Faculty of Life Sciences, University of Copenhagen, Department of Large Animal Sciences, Grønnegårdsvej 8, 1870 Frederiksberg C, Denmark; tibj@fvst.dk*

Lameness in finisher pigs affects both animal welfare and profitability. However, information about the degree of discomfort and economic losses associated with individual causes of lameness are lacking. This study quantified and compared the welfare and economic impact of 9 common causes of lameness using expert opinions. The degree of pain was used as a proxy for the level of discomfort or lack of welfare a pig suffering from each cause of lameness would experience. The specific causes of lameness included in this study were infectious arthritis caused by (1) *Mycoplasma hyosynoviae*, (2) *Erysipelothrix rhusiopathiae*, (3) *Haemophilus parasuis* and (4) *Streptococcus suis*, physical injuries such as (5) fractures of the leg, (6) lesions to the claw wall and (7) lesions to the claw sole. Additional, 2 different manifestations of osteochondrosis were considered: (8) osteochondrosis manifesta and (9) osteochondrosis dissecans. Six researchers working with animal behaviour were asked to evaluate the level of pain a pig suffering from each of the 9 causes of lameness would experience. In the evaluation an arbitrary scale from 0 (no pain) to 100 (insufferable pain) was used. Furthermore, 8 Danish pig veterinarians answered questionnaires regarding the effect of production for each cause of lameness. Hence, the probability of euthanisation, treatment with antibiotics and analgesics, and the effect on the daily weight gain and feed conversion ratio were used to calculate the resulting profit margin for a pig suffering from each of the 9 causes. To accommodate the uncertainty associated with the expert assessments, simulations were performed using the minimum, median and maximum values as input to BetaPert distributions. According to the experts, fracture caused the highest degree of animal pain and the largest reduction in productivity. Lesions to the claw wall and lesions to the volar area of the feet caused the lowest degree of pain. Among the 4 infectious pathogens included in the study, *Mycoplasma hyosynoviae* caused the lowest reduction in productivity and had a low impact on pain, whereas *Erysipelothrix rhusiopathiae* had a high impact on both animal pain and productivity. Conclusively, a more transparent evaluation of the overall impact of lameness in finisher pigs is achieved when considering the consequences of animal welfare and profitability concomitantly. More research on aggregating the consequences of lameness to the herd level is needed in the future.

Automated assessment of lameness in dairy cows with accelerometers

Chapinal, N.[1], De Passillé, A.M.[2], Pastell, M.[3], Hänninen, L.[3], Munksgaard, L.[4] and Rushen, J.[2], [1]University of British Columbia, Animal Welfare Program, 2357 Main Mall, Vancouver, BC, V6T 1Z4, Canada, [2]Agriculture and Agri-Food Canada, 6947 Highway 7, Agassiz, BC, V9M 1A0, Canada, [3] University of Helsinki, Research Centre for Animal Welfare, P.O. Box 28, Helsinki, FI-00014, Finland, [4]Aarhus University, Faculty of Agricultural Sciences, P.O. Box 50, Tjele, DK-8830, Denmark; nchapinal@yahoo.com

The prevalence of lameness is commonly used in the welfare assessments of dairy farms but detecting lameness on large farms is difficult. We examined whether measures of acceleration while cows walk could be used to automatically detect changes in gait or locomotion that were associated with lameness or differences in flooring. Acceleration measures, overall gait, visually assessed asymmetry of the steps and walking speed were assessed on cows walking on concrete (experiment 1, n=12 cows with a broad range of gait scores) or on both concrete and rubber (experiment 2, n=24 not clinically lame cows). We calculated the overall acceleration as the magnitude of the 3-dimensional acceleration vector, the variance of the acceleration and the asymmetry of variance between the front and rear pair of legs. The asymmetry of variance of the acceleration in the front and rear legs was positively correlated with overall gait and the visually assessed asymmetry of the steps ($r \geq 0.6$; $P<0.05$). Walking speed was negatively correlated with asymmetry of the steps ($r=-0.6$; $P=0.03$) and with several measures of acceleration taken in the legs and back ($r \geq 0.7$; $P<0.05$). Cows had lower gait scores (2.3 vs. 2.6; SED=0.1; $P<0.001$ measured on a 5-point scale) and lower scores for asymmetry of the steps (18.0 vs. 23.1; SED=2.2; $P=0.02$, measured in a continuous 100-unit scale) when they walked on rubber rather than concrete and their walking speed increased (1.29 vs. 1.23 m/s; SED=0.02; $P=0.03$). The difference in walking speed between concrete and rubber correlated with the difference in acceleration and the difference in the variance of the acceleration of the legs and back ($r \geq 0.6$ in all cases; $P<0.01$). Three-dimensional accelerometers seem to be a promising tool for gait assessment and lameness detection on farm and to examine suitability of flooring.

Effects of long rope tethering on dairy calf behaviour

Seo, T., Hayakawa, S. and Kashiwamura, F., Obihiro University, Obihiro, Hokkaido, 080-8555, Japan; seo@obihiro.ac.jp

EU legislation (91/629/EEC) requires that the tethering of calves be banned, except for group-housed calves, which can be tethered for no more than one hour at feeding. In Japan, there are many calves and milking cows tethered continuously at dairy farms because it is difficult for farmers having only small areas of farmland to prepare calf pens or hutches. Furthermore, some farmers tether calves to accustom them early to future tethering. The stress imparted on calves by tethering might be reduced if the calves are tethered with long ropes. This study compares the behaviour of non-tethered calves and those tethered with long ropes or short ropes. Holstein calves were housed in individual calf hutches by tethering with 140 cm lengths of long ropes (L), with 70 cm lengths of short rope (S), or non-tethered within a fenced area (1.5 × 1.2 m) (C). Five male calves were used for each treatment. The cow's ropes in L and S were tied to hutches at 60 cm height from the ground. Observations of each calf were conducted at 3 days (3D), and at 1, 2, 3, 4, 5, and 6 weeks old (1-6 wk). Examinations were made for 8 hr. Behaviour and posture were recorded using the 1 min interval sampling method. Durations of all standing up and lying down movements were recorded using continuous sampling. Comparisons were made using one-way ANOVA and post hoc least significant difference tests. The average duration of lying down was not significantly different among the calves. The standing up duration was longer in S than in C at 2 wk ($P<0.05$). Grooming, which might reflect displacement behaviour, was higher in S than L at 4 wk ($P<0.05$). Licking objects was observed more in S than in L at 3-6 wk, and more in S than in C at 4-5 wk ($P<0.05$). Lying the head on the body was observed more in S than in C at 3D ($P<0.05$), and more in S than in L at 3 wk ($P<0.05$). Additionally, lying the head on ground was observed less in S than in C or in L at 3D ($P<0.05$). These results related to the lying-down posture suggest that S calves were unable to stretch the neck forward because it was restrained by a short rope. Movement of calves tied with short rope was more restricted than those of non-tethered or long-rope tethered calves. Little difference was found between non-tied and long-rope-tethered calves. Short-tied calves apparently feel stress because their movement is restricted. Tethering of calves is not necessarily undesirable, although short rope (e.g. 70 cm) should be avoided. The calves' behaviour is not restricted compared with that of non-tethered calves if tethered with a long rope (e.g. 140 cm). Nevertheless, further research is necessary to examine the rope height from floor to ensure appropriate lying down, standing up, and lying posture.

Welfare assessment of pigs at slaughter based on critical control points

Von Borell, E. and Schäffer, D., Martin-Luther-University Halle-Wittenberg, Institute of Agricultural and Nutritional Sciences, Theodor-Lieser-Str. 11, D-06120 Halle, Germany; eberhard.vonborell@landw.uni-halle.de

Welfare assessment systems are mostly applied to the primary on farm production level. Assessment of slaughter pig handling was only done so far for some distinct processes. Therefore, a comprehensive monitoring concept has been developed and validated that allows the assessment of pre-slaughter handling under commercial conditions. This paper is focused on the assessment of unloading and moving pigs to the stunning area. Checklists with control points (CP) for all handling processes that require mandatory answers were formulated on the basis of empirical experiences from an animal welfare officer and on inspection data involving 4 millions of slaughter pigs. Handling procedures during unloading, entering into the entrance area of the CO_2-stunning (Butina® Backloader-System, Holbaek, DK) as well as the stunning success was checked within the scope of quality assurance and self-monitoring. Monitoring of unloading Control points during unloading and delivery included technical specifications of the vehicles and the unloading area, animal health, separation of stressed and sick pigs and performance of emergency slaughter. Along with the daily analysis of veterinary control data these CP could provide the basis for a supplier quality rating system. Monitoring of moving pigs to stunning The entrance area of the CO_2-stunning system and its impact on the behaviour of pigs was specifically considered as this is for welfare reasons the most critical area of concern. Development and analysis of CP was based on daily camera recordings and in situ inspections. CP included technical specifications of the chute, driving aids, movements of pigs and noise. A maximum volume of 750 pigs (in groups of 6-7) per hour and system is proposed as an upper limit to safeguard appropriate handling and stunning. We conclude that the concept of CP is suitable for the assessment of pig welfare at the abattoir.

Assessing farm animal welfare without visiting the farm – is it possible and useful?

Sørensen, J.T.[1], Houe, H.[2], Sandøe, P.[2], Otten, N.[2] and Knage-Rasmussen, K.M.[1], [1]Aarhus Faculty of Science and Technology, Department of Animal Health and Bioscience, Blichers Alle 20, DK-8830, Denmark, [2]University of Copenhagen, Faculty of Life Science, Grønnegaardsvej 8, DK-1870 Frederiksberg C, Denmark; jantind.sorensen@agrsci.dk

Animal welfare is typical assessed on farms by external observers, sometimes in the role of inspectors, doing systematic observations on animals and/or the environment. External observers are costly and efforts to minimize the time spent by external observers are giving rise to a delicate discussion about priorities between costs, validity and reliability. In this situation it is worthwhile to consider the option of systems for assessing animal welfare without an external observer having to visit the farm. Consider doing the assessment based on available data already generated for other purposes. Thus modern livestock production chains produce a range of data intended to be used for internal and external purposes, such as production management and food safety. Such data contain, often indirectly, information relevant for assessing animal welfare. Because such data are already 'paid for' in their use for other purposes, we can consider these data as 'cheap data' in the sense that they are already available. In a Danish research project, focusing on developing a cost effective on-farm welfare assessment system for the authorities, such existing data together with farmer information is examined for their information value for animal welfare compared to obtaining additional on-farm data by external observers. Data for an on-farm welfare assessment are in this study classified into four levels; (1) existing data, (2) data provided by the farmer, (3) animal based clinical observations, (4) animal behavior observations. Three livestock production systems are included; dairy cattle herds, sow herds and finisher herds. A complete welfare assessment based on level 1+2 data are in the project compared with a welfare assessment based on data from level 3+4. Unlike the other data, level 3+4 data are obtained by means of external observers. Roughly the same welfare criteria structure is used for on-farm assessment based on level 1+2 and on level 3+4. In addition to establishing the validity of data when comparing these levels the project may be instrumental in finding an answer to the following key questions: Can 1+2 data stand alone as a basis for a valid welfare assessment or are they more suitable as the basis of a screening method? Even if level 1 and 2 indicators are good predictors of poor welfare will they serve the purpose of various forms of control? If level 1+2 data are to be used for screening which parts of these data are most relevant? An overview of the findings the project these questions will be discussed.

Simplification of furnished cages for laying hens

Tanaka, T.[1], Shimmura, T.[1], Hirahara, S.[2] and Uetake, K.[1], [1]Azabu University, Sahgamihara, 252-5201, Japan, [2]Kanagawa Agricultural Technology Center, Ebina, 243-0417, Japan; tanakat@azabu-u.ac.jp

Animal welfare has been spread around the world, and various new housing systems for laying hens have been developed. In most countries, the economics and hygiene status in the housing systems are emphasized. So, furnished cage for laying hens that had the advantages in these points is one of the most notable systems. However, the furnished cages cost more than conventional cages. We proposed an idea that nest and dustbath areas were united (nest-dustbath area) to simplify the furnished cages and developed them using existed conventional cages to decrease the introduction costs. In total, 216 hens were used. In exp. 1, the following four cage designs were prepared: small conventional cages (SC), large conventional cages (LC), furnished cages with a nest-dustbath area covered all sides and a perch (FC-A), and furnished cages without perch (FC-B). Two hybrids (WL and RIR) were introduced into these cages. In exp. 2, SC, LC, FC-A and another cage design were prepared: furnished cages with a nest-dustbath area covered one side and a perch (FC-C). WL was introduced into these cage systems. Behaviours were recorded using scan sampling excluding comfort behaviours and number of steps measured by focal sampling. Physical condition, production and egg quality were also measured. After transformation, the data were analyzed using One-way ANOVA followed by Tukey-Kramer test. In both exp. 1 and 2, the proportion of hens moving, and the number of head scratching and leg stretching were higher in furnished cages than in LC, and in LC than in SC ($P<0.05$). In exp. 1, the proportion of hens moving was higher in FC-A (3.1%) than in FC-B (1.8%, $P<0.05$), while foot condition score was worse in FC-A (3.5) than in FC-B (4.0, $P<0.05$). In exp. 2, the proportions of hens performed dustbathing were higher in FC-C (1.6%) than in FC-A (0.0%, $P<0.01$). The proportions of eggs laid in the nest-dsutbath area are high and had no significant differences among the furnished cages (FC-A: 89.8%; FC-B: 87.7%; FC-C: 85.4%). No significant differences between the furnished cages were also found in the proportion of hens performed pre-laying (FC-A: 5.0%; FC-B: 5.0%; FC-C: 4.7%). There are no significant differences between cage designs in production and egg quality. As a whole, welfare level was higher in furnished cages than in conventional cages. Among furnished cages, nest-dustbath area covered one side was used as both nest box and dustbath areas, while the nest-dustbath area covered all sides was used as only nest box. These results would indicate that new-typed furnished cages that had advantages of low introduction costs, simplified resource, high welfare level and high production were developed.

Assessment of beef welfare quality® protocol in extensive systems

Huertas, S.[1], Paranhos, M.[2], Manteca, X.[3], Galindo, F.[4] and Köbric, C.[5], [1]Facultad de Veterinaria, Universidad de la República, Lasplaces 1550, 11600, Uruguay, [2]Faculde de Ciências Agrárias e Veteriárias, UNESP, Jaboticabal, 14884-000, Brazil, [3]Facultad de Veterinaria, Universidad Autónoma de Barcelona, Bellaterra, Barcelona, 08193, Spain, [4]Facultad de Medicina Veterinaria y Zootecnia, Universidad Nacional Autónoma de México, Ciudad Universitaria, 0451, Mexico, [5]Facultad de Ciencias Veterinarias y Pecuarias - Universidad de Chile, Av. Santa Rosa, La Pintana, Santiago, 11735, Chile; stellamaris32@adinet.com.uy

The European Welfare Quality® project was intended to develop standards for on-farm welfare assessment and product information systems as well as practical strategies for improving animal welfare. The system to evaluate the animal welfare was set up, however, for intensive production systems. Brazil, Chile, Mexico and Uruguay joined the WQ in 2007. The objective of the present study was to assess the applicability of the WQ to extensive and semi extensive beef cattle farms. A total of 26 beef cattle farms were visited (Brazil: Minas Gerais, São Paulo, Paraná; Chile center-south zone; Uruguay) and 4,500 animals were tested. The average size of the farms varied from 250 hectares to 2,700, and the average number of cattle per farm ranged from 752 to 3,600. The most relevant animal-based measures used were: Social Behavior: observing antagonistic and cohesive behaviour; Quality Behaviour Assessment: observing the animals and describing their emotional state; Avoidance Distance: the distance at which an animal moves away from humans and Clinical Score: which assesses physical health factors. Due to group sizes, sometimes it is impossible to use animal-based measures and in these cases resource- or management-based measures should be used. The parameters used in the scoring model would be optimized to get the overall assessment, but there is no yet gold standard measure and no available information on the relative importance animals attribute to the various welfare aspects. Results showed that in extensive systems the farm and herd size could be a problem in the protocol application and defining the appropriate sample size appeared to be one of the most difficult areas. Individual identification of the animals was difficult and doing a herd scan was not always possible. Avoidance distance found in animals grazing free on natural pastures was 34 meters on average (15-40). We conclude that more studies are needed to adapt the protocol to extensive production systems and it should include new factors such as heat stress, risk of predation, horn flies, vegetation complexity and diversity, pasture size and quality, mineral supplements, water supply from natural or man made sources, mounting and branding (compulsory in some countries and very common in others).

Radiography of laying hens: a useful tool for identifying keel bone fractures and assessing fracture healing in live birds

Richards, G.J.[1], Nasr, M.[1,2], Brown, S.N.[1], González Szamocki, E.M.[1], Murrell, J.[1], Barr, F.[1] and Wilkins, L.J.[1], [1]Univeristy of Bristol, Langford House, Langford, BS40 5DU, United Kingdom, [2]Zagazig University, Department of Animal Wealth Development, Faculty of Veterinary Medicine, Zagazig University, 44511, Egypt; g.richards@bristol.ac.uk

Bone breakage in laying hen flocks presents a considerable welfare concern within the poultry industry because of the potential pain and possible effect on behavioural patterns, associated with fractures in the live bird. The occurrence of fractures during the production cycle is primarily associated with birds housed in extensive systems because of the greater freedom of movement and thus potential to impact against objects within their environment. The most commonly damaged bone in these extensive systems is the keel which has been shown to account for up to 90% of the breaks observed. There is currently little or no published research examining bone healing in laying hens and particularly the healing of fractures of the keel bone, which is the primary site of damage in these birds. The aim of this study was to use x-ray technology to assess and characterise naturally occurring keel fractures in laying hens and monitor live birds over several weeks to examine the healing process. Twenty-four Lohmann brown commercial laying hens with varying degrees of keel fracture were used in this study. Birds were obtained from a commercial farm then housed in an experimental pen where they were closely monitored. Birds were x-rayed regularly over 6 weeks and the radiographic features and changing status of their keel fractures evaluated. The radiographic characteristics of old and new fractures were categorised and indicated that 80% of birds entering the study with 'new' fractures had healed after 35 days and five birds had incurred new breaks irrespective of their original fracture status. The technique provides valuable insight into the nature of keel bone fractures and the process of fracture healing in laying hens.

Effect of different environmental conditions in tie-stall barns on claw health in Finnish dairy cattle

Häggman, J. and Juga, J., University of Helsinki, Department of Agricultural Sciences, Koetilantie 7 (P.O. Box 28), 00014 University of Helsinki, Finland; johanna.haggman@helsinki.fi

Claw disorders are a major welfare problem in dairy farming. Many claw disorders are painful for the cow and claw health should be taken into account when talking about dairy cattle welfare and should be included to welfare assessment schemes. In 2010 77.4% of the Finnish dairy barns were tie-stall barns. Data from 18 038 Ayrshire and Holstein cows in 609 tie stall herds was used to analyse how different environmental conditions in farms affect on the prevalence of claw disorders. The data was collected by hoof trimmers between years 2005 and 2009. Claw disorders included to analyses were sole haemorrhage, interdigital dermatitis, sole ulcers, white-line disease, heel horn erosion, corkscrew claw, chronic laminitis and digital dermatitis. Data was analyzed using R statistical software package. A logistic generalized linear model with hoof-trimmer and farm (within hoof-trimmer) as random effects were fit to dataset. Breed had a large effect on the prevalence of claw disorders; 43.1% of Holstein and 32.4% of Ayrshire cows had claw disorders. Claw disorders increased together with parity. Cows in parity >3 had 5.19 times bigger risk to have claw disorders compared to primiparous cows. There were differences in the prevalence of claw disorders on the farms having different concentrate feeding systems. Farms using flat rate feeding (OR=1.49, P<0.001) or partial mixed ration (OR=3.68, P<0.01) had more claw disorders than farms which adjusted feeding according to yield. Cows which were held in barns with solid manure handling system with long-stand stalls had 1.37 times (P<0.001) bigger risk to have claw disorders than cows which were held in barns which had slurry manure system. Cows which had rubber mats/mattresses in cubicles had less claw disorders than cows standing on concrete (OR=0.74, P<0.001). No significant differences were found between different bedding materials. Cows which were in pasture in summer and exercise yard in winter had less claw disorders than cows which were always inside (OR=1.28, P<0.001). According to these results it can be recommended for accomplishing better claw health that farmers adjust feeding according to yield and avoid barns with solid manure handling system with long-stand stalls. It is also essential that cows have access to pasture or outside exercise yard at least in summer time.

Health and behavioral measures in farmed foxes: inter-observer reliability of farm averages

Ahola, L., Koistinen, T. and Mononen, J., University of Eastern Finland, Department of Biosciences, P.O. Box 1627, 70211 Kuopio, Finland; leena.ahola@uef.fi

WelFur project aims at developing Welfare Quality®-like on-farm welfare assessment protocols for farmed fur animals. For foxes, a preliminary protocol with potential health and behavioral measures was developed in 2010. The aim of the present study was to assess inter-observer reliability (IOR) of some of these potential measures at farm level. In autumn 2010, three assessors visited altogether 18 commercial Finnish fox farms and assessed foxes' health and behavior. The median number of foxes per farm was 3025. The number of foxes observed on the farms varied according to the total number of animals on the farm, averaging 109 and 149 foxes per farm for health and behavioral measures, respectively. On 12 out of 18 farms, the assessors observed the same individual animals. On the six remaining farms, all assessors observed different animals. All three assessors recorded the health of the foxes whereas the foxes' behavior was observed by only two assessors. The results were calculated for each assessor as '% of foxes on the farm without any signs of the problem in question' and '% of foxes on the farm expressing the behavior in question'. IORs of the calculated farm percentages of the health and behavioral measures were analyzed with Kendall Correlation Coefficient W and Spearman rank correlation, respectively. Taking into account all 18 farms, IOR at farm level was high (Kendall's $W \geq 0.65$) for four (obesity, fur chewing, moving difficulties, eye inflammation) and moderate ($0.35 \leq W < 0.65$) for six (skin lesions, foreleg weakness, toe and paw problems, dirtiness of fur, impaired mouth and teeth health, obviously sick) out of the ten health measures. When the assessors observed the same individuals, IOR was high ($W \geq 0.65$) for six out of the ten health measures, but only for four measures when the assessors observed different animals on the farms. IORs at farm level for three (active, on the platform, resting) out of the ten behavioral categories were high ($r \geq 0.65$) when taking into account all 18 farms or the 12 farms where the assessors assessed the same individual foxes. When the assessors observed different individuals on the farms, IORs at farm level for these behavioral categories were only moderate (for all $r < 0.65$). IORs at farm level were low or could not be assessed for most of the behavioral categories (e.g. stereotyped behavior) due to their rare occurrence. In conclusion, results from on-farm measures may vary depending on the animals that are being observed, especially in case of behavioral measures. In order to get more reliable results at farm level, measure descriptions in the protocol should be refined, and measures should be taken from a larger number of animals on each farm.

Inter- and intra-observer reliability for assessing sheep welfare on pasture

Mialon, M.M.[1], Brule, A.[2], Gaborit, M.[1], Davoine, J.M.[2,3], Ribaud, D.[2], Boivin, X.[1] and Boissy, A.[1], [1]INRA, URH1213 Theix, 63122 St Genes, France, [2]Institut de l Elevage, Monvoisin, BP 85225, 35652 Le Rheu, France, [3]Fédération des Alpages d Isère, La Grange, 38190 Les Adrets, France; Anne.Brule@inst-elevage.asso.fr

Farm animal welfare is a societal concern and there is a need for an overall assessment at farm level. Until now, such assessment has been more developed for intensively reared farm species than for extensively reared ones such as sheep. The aim of our study was to validate an approach to assess welfare in sheep on pasture by estimating inter- and intra-observer reliability. Since welfare is a multidimensional concept, we proposed animal-based measures from the four principles of WelfareQuality® method: good feeding, housing and health, and appropriate behaviour. Ten experimental farms of Romane ewes were used. The measures were simultaneously recorded by two trained observers and were (1) at individual level on 30 ewes and (2) at group level on 3×30-ewes groups. Ewes were randomly selected from the flock. All the measures were collected twice on two consecutive days. The individual measures concerned the first three principles (body condition score, animal cleanliness, wool humidity, lameness, lesions, respiratory disorders and hoof overgrowth). The appropriate behaviours were assessed from several group measures. First, overall reactions to a novel object (a black box) and then to a sudden event (a plastic bag blew out from the box) were recorded during 40 min by scan sampling with a 5-min interval before and a 30-sec interval after the sudden event. Second, overall reactions to the approach by an unfamiliar human (one of the observers walked toward the sheep at a regular pace) were recorded: distance between the human and the front of the group, and between the front of the group and the bottom of the laneway, and time taken to pass the human. The inter- and intra-observer reliability was assessed using intra-class correlation coefficients for quantitative data analysed with a mixed model and kappa coefficients for ordinal qualitative data. Inter-observer reliability was good for individual measures as well as for group measures (>0.66). Intra-observer reliability was good for individual measures (>0.65) except for wool humidity (<0.1) but was poor (<0.2) to moderate (<0.6) for group measures. Changes in environmental factors and/or familiarization effect due to the test repetition could partly explain such poor intra-observer reliability for behavioural measures. Therefore, our preliminary results suggest that individual physical criteria appear much more reliable than behavioural ones in order to evaluate sheep welfare on pasture. A further study will estimate the variability of indicators between various commercial sheep farms from our approach.

Assessing cow comfort with variable weight for parameters depending on their score

Van Eerdenburg, F.[1]*, Sossidou, E.*[2]*, Saltijeral-Oaxaca, J.*[3] *and Vázquez-Flores, S.*[4]*,* [1]*Fac Veterinary Medicine, Farm Animal Health, Yalelaan 7, 3584 CL Utrecht, Netherlands,* [2]*N.AG.RE.F, 19 Egialias & Chalepa, 15125 Maroussi, Greece,* [3]*Univ Aut Metropol, Calz del Hueso 1100, 04960 México D.F, Mexico,* [4]*ESIABA, Tecnológico de Monterrey-campus Querétaro, Epigmenio Gonzalez 500, 76130 Querétaro, Mexico; F.J.C.M.vaneerdenburg@uu.nl*

The aim of this study was to develop a scoring system for cow comfort using a set of animal indicators. The system assesses cow and environmental parameters, and depending on the score, it has a variable weight for all parameters. Applicability in practice has been leading during the development and resulted in a system that can be executed in less than one hour. The scoring is based on available reports and experience of the authors and was extensively validated in practice over three years. It is constructed of several chapters, which need each to score a minimum number of points. If not, the difference between the score and the minimum is subtracted from the total score and thus, increasing the weight of this chapter in the total score. The fact that negative scores weigh more than positive ones, is unique for this system. If a certain aspect of welfare, e.g. food, is negatively scored, this implies that there is an urgent need for that particular aspect. If an animal is hungry, the search for food is dominating over other needs, like proper bedding or social contact. Conversely, a well fed animal tend to have more social contact or rest properly. Trained investigators visited farms in three countries: The Netherlands (67), Mexico (55) and Greece (8). Because of the different climatic conditions in the three countries, the data were treated separately. The score had a positive correlation with the Welfare Quality® system of 0.4 ($P<0.05$) in the Dutch study and 0.8 ($P<0.05$) in Greece. The milk yield was positively correlated with the total score in the Mexican study ($r=0.13$; $P=0.35$) and in the Dutch study ($r=0.38$; $P=0.04$). This is an important aspect for the promotion of a better welfare for dairy cattle to the farmers. The scoring system was used by many persons and on many farms. It was concluded that after a short training all observers could evaluate a farm in less than 1 hour, if the farmer had the historical health data ready. So it is a system that can be implemented in the routine of herd health consultants. This in contrast to the Welfare Quality® system, that takes approximately 7 h for a 150 cow farm after an extensive training. Besides a score, the system developed, provides the herd health consultant with an overview of the areas that need attention. Because it is numerical, one can compare the comfort level of the cows between farms, world wide.

Effect of summer grazing on the probability of hock joint injuries in intensive dairy cow production systems

Burow, E., Rousing, T., Thomsen, P.T. and Sørensen, J.T., Aarhus University, Faculty of Agricultural Sciences, Animal Health and Biosciences, Blichers Allé 20, 8830 Tjele, Denmark; Elke.Burow@agrsci.dk

Existing animal welfare assessments do not yet consider the possible effect of grazing on the welfare of dairy cows. However, a commonly used welfare indicator as alterations of the integument is linked to the cows' environment. Injuries like e.g. swellings or wounds can lead to pain and aversive emotions of dairy cows and may in general be seen as a problem by the public. Our aim was to investigate possible effect of grazing (here among different amount of grazing time) on the hock joint integument in summer grazing herds during winter and summer season as well as in zero-grazing herds. We hypothesized that the probability of alterations at the hock joint integument in summer grazing herds would (a) be lower in summer season compared to winter season, (b) be decreased by an increase in time spent on grass and (c) be lower than in zero-grazing herds. Dairy cows from 35 Danish summer grazing herds (cubicle housing system, average herd size 164±70 cows) were examined once in winter and once in summer during the grazing period in 2010. In addition, 21 zero-grazing herds (cubicle housing system, average herd size 173±60) were visited once in 2010. A random sample of in average 63±7.6 cows, sized in relation to the individual herd size, was selected by systematic random sampling before each visit. As part of the applied comprehensive welfare protocol (modified after Welfare Quality®), the hock joint integument of each sampled cow was assessed. The hock joint integument was either scored 0: for no alterations or hairless patches smaller than 2 cm in diameter, (1) for at least one hairless patch of minimum 2 cm in diameter or (2) for at least one wound or swelling. In the 35 grazing herds, daily grazing hours were noted during one month before clinical assessment. The mean daily grazing hours was grouped into two levels: 1: 3-9 hours (18 herds), 2: >9-21 hours (17 herds). Additionally, the surface material of the cubicles was noted and categorized as (1) rubber mattress (25 grazing and 17 zero-grazing herds) or (2) straw, chipped wood, sand or turf (8 and 4 herds). Results of preliminary analyses show that grazing herds in winter season had in median 77% and in summer season 42% cows with a hock joint scored as either 1 or 2. The 21 zero-grazing herds had in median 66% cows with these scores. These preliminary results indicate an improvement of hock joint alterations of the cows during the grazing season. Additional statistical analyses are to be carried out and will be presented at the conference.

A new model for corporate-driven animal welfare assessments: benchmarking cow comfort as a service for dairy farmers

Von Keyserlingk, M.A.G.[1], Galo, E.[2], Gable, S.[2] and Weary, D.M.[1], [1]University of British Columbia, Animal Welfare Program, 2357 Main Mall, Vancouver, BC, V6T 1Z6, Canada, [2]Novus International Inc., 20 Research Park Drive, St. Charles, MO, 63304, USA; nina@mail.ubc.ca

Corporations have played an important role in developing and delivering on-farm animal welfare assessment schemes, but to date these have been focused on evaluating whether or not individual farms meet the corporation's buying standards. Here we describe a novel approach to corporate involvement in on-farm animal welfare assessment, driven by the desire to provide a service for farmers and a vehicle for engagement on issues of cow management. Novus International Inc., a company with an animal nutrition focus, launched the COWS benchmarking project. This program, focused on animal-based measures (including gait score, hock and knee injuries, and lying time and number of lying bouts) and facility-based measures (including aspects of free stall design, feeder design and management and other management practices including stocking density, and bedding practices). The program was originally developed by the University of British Columbia and piloted on 43 farms in British Columbia. The COWS program has since been implemented on approximately 150 dairy farms in California, New York, Pennsylvania, Vermont, Texas and New Mexico. The principle objective has been to provide individual farms with data from their own farm and averages from other farms in their region that they can use to benchmark their own performance. Farmers are provided confidential reports, and these reports are used as a basis for a discussion between the farmer, a Novus employee, and others involved in the management of that farm such as the nutritionist and herd veterinarian. The aim of this discussion is to identify changes in farm design and management that can be used to address concerns emerging from the report. Farmers that have participated have especially valued the detailed reports on their own farms, allowing them to make well-informed decisions and develop tailored strategies for improving the care and management of cows on their farm. For Novus, this program has provided the opportunity for one-on-one engagement with producers, and the halo effect of being associated with a valued service. These benefits have led to the recent decision by Novus to expand the COWS program, helping to meet the many requests we continue to receive from farmers interested in participating in this benchmarking exercise.

Evaluating the diagnostic performance of indicators of young lamb welfare

Phythian, C.J.[1], Toft, N.[2], Michalopoulou, E.[1], Cripps, P.J.[1], Grove-White, D.[1] and Duncan, J.S.[1], [1]University of Liverpool, Leahurst, CH64 7TE, United Kingdom, [2]University of Copenhagen, Grønnegårdsvej, 8 DK-1870 Frederiksberg C, Denmark; clare.phythian@googlemail.com

Scientifically valid, robust and transparent indicators are needed by the sheep industry, farm certification, and enforcement agencies to assess on-farm standards of lamb welfare. Currently, indicators for the on-farm assessment of lamb welfare conforming to these criteria do not exist. Therefore, the objective of this study was to develop valid, reliable and feasible indicators for the on-farm welfare assessment of young lambs (aged ≤6 weeks). A literature review and consensus of expert opinion identified a diverse number of on-farm welfare concerns for young lambs (n=53) including starvation, hypothermia, and the presence of conditions such as lameness and ocular abnormalities. Consequently, 4 non-invasive welfare indicators: demeanour, lameness, body condition and eye condition were developed by assessing the behaviour and physical appearance of individual lambs. The indicators were assessed by 4 trained observers – 2 veterinary surgeons (A and C) and 2 animal-science students (B and D) on a total of 966 young lambs on 17 farms. Inter-observer reliability was examined using Fleiss's κ and graphical distributions of scoring differences. Latent Class Analysis (LCA) estimated the diagnostic sensitivity (Se) and specificity (Sp) of each observer and predicted the test performance of unknown, random observers who may perform future indicator assessments. The assessment of demeanour produced a κ 0.55, Se of 0.75 – 0.85 and consistently high Sp (0.98 – 1.00). Eye condition assessments were also consistent across observers (κ 0.72, Se 0.86 – 0.89, Sp ≥0.99). Lameness assessments achieved a κ 0.68, Se ≥ 0.7 and Sp 1.00 for lameness assessment. Body condition produced κ 0.71 and observers A, B and D achieved higher levels of test performance (Se ≥0.80, Sp ≥0.99) than observer C (Se 0.38, Sp 0.98). In addition, few scoring disagreements occurred. The predicted performance of future observers also appeared promising: demeanour (Se 0.78, Sp 0.98), eye condition (Se 0.88, Sp 0.99), body condition (Se 0.74, Sp 0.99) and lameness (Se 0.76, Sp 1.00). Demeanour, lameness, body condition and eye condition achieved good levels of diagnostic performance when applied by a group of trained observers. LCA provided a useful means of evaluating the test validity of animal-based welfare indicators and it is suggested that unknown observers may achieve similar levels of Se and Sp if these indicators are applied in future on-farm welfare assessments of young lambs.

Evaluating the diagnostic performance of indicators of sheep welfare

Phythian, C.J.[1], Toft, N.[2], Michalopoulou, E.[1], Cripps, P.J.[1], Grove-White, D.[1] and Duncan, J.S.[1],
[1]University of Liverpool, Leahurst, CH64 7TE, United Kingdom, [2]University of Copenhagen, Grønnegårdsvej, DK-1870 Frederiksberg C, Denmark; clare.phythian@googlemail.com

Sheep farmers and veterinary surgeons frequently judge the standard of flock welfare by assessing physical measures of health and welfare, such as the body condition or gait, of individual sheep. Some of these measures require examination of the individual sheep and are akin to diagnostic tests that are performed during clinical assessments. The aim of this study was to apply methods used to evaluate the validity of diagnostics as a means of examining the diagnostic performance of animal-based indicators of sheep welfare. Eight observers independently tested the individual animal indicators on 1,146 adult sheep from 38 flocks to evaluate the level of inter-observer reliability, sensitivity (Se) and specificity (Sp) of observer assessments. The overall level of inter-observer reliability was evaluated by Fleiss's kappa (κ). Latent class analysis (LCA) determined the diagnostic Se and Sp of each observer and also predicted the test performance of unknown observers who may apply these indicators in the future. Assessments of demeanour, mastitis and wool loss achieved $\kappa>0.75$, and examination of tooth condition, pruritus and lameness and body condition produced κ values $>0.41<0.74$. LCA found that all observers had high levels of diagnostic Sp (≥0.97) for all indicator assessments. High Se (≥0.98) was found for the assessment of demeanour and wool loss and with the exception of one observer, good levels of Se were found for body condition scoring (≥0.95). Se ranged 0.86-0.91 for lameness and 0.50-0.72 for mastitis assessments. Lower levels of Se were found for examination of pruritus (0.13-0.40). With the exception of pruritus, LCA predicted that unknown observers could possibly produce high levels of Sp and good Se if performing assessments of the 8 animal-based indicators of sheep welfare in the future. The application of diagnostic test principles to the evaluation of animal-based indicators of sheep welfare found that observers were reliable ($\kappa>0.41$) at identifying specific conditions, such as mastitis or lameness, in individual sheep. The low number of sheep in the study population with pruritic skin conditions was suggested to produce the lower level of Se identified in this study. Therefore, it is suggested that the test performance of these indicators could possibly be considerably higher if the indicators were tested on a population with a higher proportion of conditions associated with poor sheep welfare.

Heat tolerance in crossbred sheep

Lima, F.G.[1], Costa, G.L.[1], Ribeiro, C.S.[1], Oliveira, N.A.[1], Cardoso, D.[1], Laudares, K.M.[1], Vaz-Jr, R.[1], Assunção, P.[1], Cardoso, C.[2], Louvandini, H.[3], Fioravanti, M.C.S.[1] and Mcmanus, C.M.[4], [1]Univ. Federal de Goiás, C. Samambaia, 74001970, Goiânia-GO, Brazil, [2]Univ. de Brasília, C. Darcy Ribeiro, 70910970, Brasília-DF, Brazil, [3]Univ. de São Paulo, Av. Centenário, 13400970, Piracicaba-SP, Brazil, [4]Univ. Federal do Rio Grande do Sul, Av. Bento Gonçalves, 91540000, Porto Alegre-RS, Brazil; fglfgl@hotmail.com

The sheep industry has increased greatly in the central west of Brazil. One of the main challenges faced by producers is excessive heat and low humidity in the region during some seasons. An alternative is the cross-breeding for heat-tolerant animals that suffer less from the typical climate of the region. A total of 48 lambs aged six months were divided into eight genetic groups. The crosses were: 50%East Friesian × 50%Santa Inês (G1), 50%Primera × 50%Santa Inês (G2), 87.5%Poll Dorset × 12.5%Santa Inês (G3), 100%Santa Inês (G4), 50%Poll Dorset × 50%Dorper (G5), 50%Poll Dorset × 50%Santa Inês (G6), 50%Poll Dorset × 50%White Dorper (G7), 75%Poll Dorset × 25%Santa Inês (G8). Heart (HR), respiratory rates (RR), rectal temperature (RT) and superficial temperature were measured in animals and ground at 7:30 AM and 11:30 AM during three days. To obtain the surface temperatures of animals and the ground, thermographic images were made with an infrared camera (FLIR i-Series®) with QuickReport® software for data collection. Statistical analysis was performed using the Statistical Analysis System®, evaluating the effect of temperatures from the ground, the environment and the black globe and humidity, wind speed, and genetic group on temperatures and physiological measures of animals. During the three days of the trial at 7:30 AM the average temperature recorded was 23.7 °C, humidity was 71% and wind speed 0 m/s, and black globe temperatures was 27.3 °C in the sun and 22.8 °C in the shade. At 11:30 AM the average temperature recorded was 30.7 °C, humidity was 46% and wind speed 0 m/s, and black globe temperatures was of 48.2 °C in the sun and 30.8 °C in the shade. The G5 was less adapted to environmental conditions, where 100% of the lambs showed RT above 39.9 °C and had higher rates of HR and RR at 11:30 AM, suggesting that the animals had difficulty maintaining homeothermy. The G2 was more adapted, where only 18.75% showed RT above 39.9 °C at 11:30 AM, and also had lower levels of HR and RR. The infrared thermographic images showed that the G3 was also well adapted. This group maintained lower surface temperature than the other groups. In evaluating of heat tolerance by Rauschenbach Yerokhin test, the G4 was better adapted to the environment. The assessment of all parameters showed that certain genetic groups suffered severe thermal stress and others tolerated the heat.

Assessment of animal welfare on dairy farms in Santa Catarina state, Brazil

Cardoso Costa, J.H., Hotzel, M.J., Machado Filho, L.C.P., Balcão, L.F., Dáros, R.R. and Bertoli, F., LETA – Laboratório de Etologia Aplicada, Universidade Federal de Santa Catarina, Rod. Admar Gonzaga, 1346 – Itacorubi, 88034-001, Florianópolis, SC, Brazil; juaohcc@gmail.com

A holistic approach was used to assess the welfare of dairy herds in the west of Santa Catarina, South Brazil. Data were colleted in the spring and summer months of 2009-2010 from 120 family-run dairy farms, distributed in 24 municipalities that produce around 20% of the milk of the state. Average herd size was 24.6±18.9 cows (5 – 111 cows); total daily milk production was 423.3±530.3 l (50-3,320 l). All farms used pasture-based systems, with varying levels of supplementation with concentrates and silage. A questionnaire was followed by inspection of the production environment and of the animals, to assess aspects of the living environment – the milking parlor, the calf barn, the grazing and waiting areas used by the animals-, management and health of the animals. Automatic milking was used in 98% of the farms. During milking, 71% of the farmers tied the cows. Veterinary procedures were carried out in the milking parlour in 35% of the farms and in the feed bunk in 48%. Pets were present in the milking parlour in 46% of the farms, and other farm animals in 37%. Clinical or subclinical mastitis were diagnosed in 30% of the cows; 27% of cows were infested by ectoparasites and 4% were lame. On inspection, 35% of the lactating cows had a soiled side, 70% had dirty legs and 20% had dirty udders. Average body condition score was 2.8±0.5 (1.5-4.5 in different herds); 8% of the cows had a body condition score 2 or lower. Hock lesions were found in 14% of the cows, hipbone or tailbone lesions in 4%. Average flight distance, assessed in the paddock, was 3.0±1.7 m. Water was available ad libitum in 42% of the farms, whereas 55% of the herds had access to water only in the resting areas and in the milking parlour; 3% of the herds had drinking water available only in the milking parlour. In 68% of the farms the herd was moved to paddocks with shade during the hottest hours of the day; only 16% of the herds had permanent access to shade and 16% did not have any shade available. Unweaned calves were housed individually in 66% of the farms. They received 3 l (9% of the farms), 4 l (67% of the farms), or more than 4 l (13% of the farms) of milk for 64.8±23.7 days – though 7% the calves were weaned at 30 days of age or less. Overall, the results of this survey highlight the need for changes in environmental design and management practices in order to improve the welfare and productivity of dairy herds. This data may help guide future research and extension projects to improve these issues.

Risk based animal welfare assessment in sow herds based on central database information on medication

Knage-Rasmussen, K.M.[1], Houe, H.[2], Rousing, T.[1] and Sørensen, J.T.[1], [1]Faculty of agricultural science, Department of animal health and bioscience, Blichers Allé 20, Postboks 50, DK-8830 Tjele, Denmark, [2]Faculty of Life science, Department of large animal science, Bülowsvej 17, 1870 Frederiksberg C, Denmark; Kristian.Knage-Rasmussen@agrsci.dk

Animal welfare assessment systems – such as Welfare Quality® – are based on animal based welfare indicators as well as on farm resource measures the collection of which is rather cost intensive. There is a need for less expensive welfare indicators by the use of information from central and official databases. The objective of this paper is to quantify the use of medicine recorded in a central database as well as information on housing system and to explore how well this information reflect the 'true welfare state' of the animals. The paper will focus on the use of analgesic medication in sow herds and will compare these measures with a clinical on farm evaluation. Forty farms agreed to participate in this study. Information on medication was obtained from the central database Vetstat, which records all medication used in Denmark. In this central and official database, the use of and handling out prescription-only medication for fur and food animal production is recorded by the veterinarians and pharmacies and further by feed mills when handing out licensed food prescription-only medication and coccidiostatic medication. The use of analgesic in sow herd is mainly ordered to the diagnostic groups 'Joints, limbs, hooves, CNS and skin'. The information is on herd level for each medication and specified for animal species, age group and diagnostic group. The information on housing system was been collected through a telephone interview and the clinical scores were assessed on farm as a random sample of gestation sows. The paper will focus on 'lameness scores' as an example of a gold standard for the animal welfare status. The farms were divided into three types of gestation housing systems: (1) Loose housing (loose housing throughout the gestations period), (2) Crates + Loose (Crates until 4 weeks after mating/insemination, loose housing in the remaining gestation period), and (3) Crates (Crates throughout the gestation period). A descriptive analysis will be completed as well as estimation of the sensitivity and specificity of using analgesic data and housing data to identify sow herds with risk of having welfare problems such as lameness. The data are at present being analysed.

Effects of milkers' attitudes and behavior on cows' avoidance distance and impacts on udder health in Swiss dairy herds

Ivemeyer, S.[1]*, Knierim, U.*[2] *and Waiblinger, S.*[3]*,* [1]*Research Institute of Organic Agriculture, Animal Health Division, Ackerstrasse, 5070 Frick, Switzerland,* [2]*University of Kassel, Department of Farm Animal Behaviour and Husbandry, Nordbahnhofstraße 1a, 37213 Witzenhausen, Germany,* [3]*University of Veterinary Medicine, Institute of Animal Husbandry and Animal Welfare, Veterinärplatz 1, 1210 Vienna, Austria; silvia.ivemeyer@fibl.org*

In a cross-sectional study we investigated effects of milkers' attitudes and behavior on cows' behavior as well as their impacts on udder health while also considering herd management factors. All 46 investigated Swiss dairy herds were kept in loose housing systems. All farms participated in an extension program for preventive mastitis control. Milkers' attitudes were assessed by a questionnaire. Milkers' acoustic and tactile behavior as well as cow behavior were observed during milking. The cows' avoidance distances in the barn towards an unknown person were recorded. Furthermore, herd management factors were assessed by questionnaire guided interviews and observations, respectively. Udder health was evaluated using the indicators prevalence of quarters with elevated somatic cell counts (>100,000 cells/ml) and prevalence of mastitis quarters (>100,000 cells/ml and culturally positive) calculated from quarter-milk-samples of all lactating cows at the time of assessment. After univariable pre-selection of associated factors, multivariable linear regression models with stepwise backwards elimination of factors with $P \geq 0.05$ were calculated on herd level. Lower cows' avoidance distances were associated with (1) positive milkers' attitude concerning importance of contact with their animals, (2) all milkers knowing all cows individually, (3) breeding selection on manageability, (4) generous dimensions of cows' lying places and (5) longer contact of the stockpersons with the animals during routine work. Predictors for higher prevalences of quarters with elevated somatic cell count were a lower percentage of positive interactions of milkers out of all interactions with the cows during milking ($P=0.030$) and a higher amount of fearful cows in the herd (with an avoidance distance above 1 m; $P=0.014$). Higher prevalences of mastitis quarters were associated with (1) again a lower percentage of positive interactions of milkers during milking ($P=0.033$), (2) breed, especially Holstein in comparison to Swiss Fleckvieh ($P<0.001$), (3) higher lactation number ($P=0.020$) and (4) generous dimensions of lying places ($P<0.001$). In conclusion, human-animal-relationship was found to be relevant for udder health and should get more attention as one possible action point in extension programs for preventive mastitis control.

Challenges in using Welfare Quality® principles for the development of an on-farm welfare assessment system for mink

Møller, S.H. and Hansen, S.W., Aarhus University, Animal Science, P.O. Box 50, 8830 Tjele, Denmark; steenh.moller@agrsci.dk

In the Welfare Quality® (WQ) on-farm welfare assessment system for cattle, pig, and poultry, animal welfare measures can be taken at almost any time of the year. This is different in the WelFur project using WQ principles to develop welfare assessment systems for mink and foxes. These species are seasonally synchronised in their strict annual cycle of production. In mink production, breeders are present from pelting to mating (December to March), breeders and kits from gestation to weaning (April to July) and breeders and growing kits from separation to pelting (July to November). Most animal-based measures can only be taken with high validity, reliability and feasibility during a few months each year: January 1st to February 24th (Period 1), May 15th to July 15th (Period 2) and October 1st to November 30th (Period 3). An advantage of this seasonality is that the measurement of welfare can be optimised and standardised in terms of age/season and sample size, making reliable results relatively cheap to obtain. A disadvantage is that measures must be taken at all farms within the two months for each period. Some welfare measures are relevant in only one period, others in two or all three periods. This presents a special challenge to the integration of data because all periods take place on the same farms, while cattle, pig, and poultry production systems are usually specialised, e.g. in reproduction (sows and piglets), growth (veal calves, fattening cattle, growing and finishing pigs, broilers) or other parts of production (dairy cows, laying hens). WQ is based on measures taken by an independent external audit on a one day farm visit. This constraint limits the use of time series data to those that all farms must have, e.g. mortality data, but excludes others on e.g. body condition development that the farmer may provide and which has been used in previous welfare assessment systems. The WelFur assessment system must address a wide range of climatic conditions because mink are housed in more or less open sheds under ambient temperatures. Cooling systems may thus only be relevant in some parts of Europe while freeze protection of the watering system is only relevant in other parts of Europe. Under these constraints 11 animal-based, 9 resource-based and 5 management-based measures have been selected. Of the four WQ principles 'Good housing' is not covered by an animal-based measure, and 7 of the 12 criteria do not include animal-based measures. The WelFur protocol for period 1 has been tested on 9 mink farms in late February, and the protocol for period 2 will be tested in June. Results on body condition and stereotypy will be presented at the conference.

Stereotypic behaviour: a useful indicator for unfulfilled feeding motivation in mink

Hansen, S.W., Damgaard, B.M. and Møller, S.H., Aarhus University, Animal Science, Blichers Allé 20, Tjele, 8830, Denmark; steffenw.hansen@agrsci.dk

Mink chosen for breeding are slimmed during the winter and flushed just before mating. However, the slimming procedure may increase the development of abnormal behaviour such as stereotypy. Stereotypic behaviour in mink is primarily observed prior to the normal feeding time during the winter period and therefore we hypothesized that the occurrence of stereotypy in winter is mainly a reflection of the feeding motivation of the mink. In order to investigate this we compared the level of feed allowance and stereotypies in 784 female mink during the winter period. The stereotypic behaviour was registered by scanning observations on Tuesdays and Thursdays in weeks 2, 4, 5, 8, and 13 in 2009. On Tuesdays, stereotypies were registered once an hour from sunrise to sunset. On Thursdays the feeding time was postponed from 11.00 h to 13.00 h and stereotypies were registered once an hour from 9.00 h to 12.00 h. It took approximately 45 minutes to complete the observations. Furthermore, we tested whether a larger allowance of low energy feed in week 4-6 could reduce the feeling of hunger and thereby decrease the performance of stereotypies. Stereotypies were almost exclusively observed 1-2 hours before feeding time (11.00 h). During postponed feeding the occurrence of stereotypies were seen 1-2 hours before expected feeding time and remained at an elevated level for 1-2 hours after expected feeding time. The occurrence of stereotypies during daytime observations was affected by loose of bodyweight ($F_{1,750}$=46.18; P=0.0001) and the greater the weight loss the more stereotypy was observed. Stereotypy was almost absent in week 2 (0.06%) but significantly increased over time, to 4.3%, 4.4%, and 15.9% in week 4, 6, and 8 and decreased again to 3.3% in week 13 after flushing. The type of feed (standard vs. low energy feed) in week 4-6 decreased the occurrence of stereotypy in week 4 (P=0.0294) and 6 (P=0.0895) but did not have lasting effects in week 8 and 13.The occurrence of stereotypy during postponed feeding reflected the level of stereotypy during the daytime observations. Based on the correlation between stereotypy and feed restriction, the hypothesis that stereotypy in winter is mainly a reflection of feeding motivation in mink was accepted. The timing of stereotyped behaviour in relation to the feeding time, and the relationship between the level of stereotypy and the degree of feed restriction, makes stereotypic behaviour a useful indicator of unfulfilled feeding motivation during the winter – clearly important for welfare – in farmed mink.

Dystocial dairy calves: condemned to poor welfare?

Barrier, A.C., Haskell, M.J. and Dwyer, C.M., SAC, Edinburgh, EH93JG, United Kingdom; alice.barrier@sac.ac.uk

Prenatal and perinatal stress can have long-term repercussions on the welfare of a young animal. In various species, the early experience of a difficult birth is traumatic and has adverse effects on the newborn. However, studies on calves have been mostly limited to beef breeds and to the neonatal period. The objective of the studies presented was to investigate the effects of birth difficulty on the welfare of dairy calves. In the first study, records from Holstein calves born on our experimental farm between 1990 and 2001 (n=2,272) were retrieved along with their birth difficulty scores (N: no assistance; FN/FM: farm assistance without/with malpresented calf; V: vet assistance) and the occurrence of stillbirth. For the liveborn heifers (n=1,237), growth rate to weaning (n=1,151), age at first service (n=1,011) and at first calving (n=796) were analysed using REML after grouping FM and V scores together (MV). Their survival to weaning, 120 days, first service and first calving was analysed using survival analysis. Stillbirth rates were 6 and 7 times higher in FN and MV calves, respectively, compared to N calves ($P<0.001$). In liveborn dystocial heifers, there was no evidence of an impaired growth to weaning or subsequent fertility ($P>0.05$). However, compared to N calves, FN calves were about three times more likely to have died by weaning, by 120 days and by first service ($P<0.05$) but significance disappeared by first calving ($P>0.05$). In a second study, we monitored a cohort of calves born from Sept 2008 to Aug 2010 on our experimental farm from birth to weaning (N: n=378; FN: n=93; FM: n=18; V: n=7). All but one V calves were born dead. FN and FM calves were 2.7 and 7 times more likely to be born dead than N calves. At weaning, mortality rates in FN liveborn heifers were 2.3 times higher than N calves, and 40% of the FM heifers had died. In their first 24 hours of life, FN and FM calves had higher median salivary cortisol levels compared to calves born normally (FN: x1.6; FM: x4; $P<0.001$) which may indicate that they experienced higher physiological stress. Serum immunoglobulins as estimated by Zinc Sulphate Turbidity tests in their first week of life were also lower in FN calves (-40%; $P=0.03$), which put them at higher risk of contracting diseases. To conclude, when they survive the birth process, dystocial calves experience lower immunity, higher mortality and likely higher physiological stress. In the dystocial calves that survive, absence of effects on growth to weaning and fertility may be explained by the mortality of the most badly affected calves or by farm management. We suggest that dystocial dairy calves have poorer welfare in the neonatal period and possibly beyond. Strategies should be implemented to improve their welfare and to lower the occurrence of dystocia.

Assessment of fearfulness, stress and feather damage in commercial laying hen parent stock flocks

De Haas, E.N.[1], Ten Napel, J.[2] and Rodenburg, T.B.[3], [1]Wageningen University, Adaptation Physiology Group, Marijkeweg 40, 6700AH, Wageningen, Netherlands, [2]Wageningen Livestock Research Centre, Wageningen University, Animal Breeding and Genomics Centre, Postbus 65, 8200AB, Lelystad, Netherlands, [3]Wageningen University, Animal Breeding and Genomics Centre, Marijkeweg 40, 6700AH, Wageningen, Netherlands; elske.dehaas@wur.nl

Laying hen parent stock farms can vary in flock size, stocking density and management practices, which can affect birds' ability to cope with fear and stress and their propensity to develop feather pecking. Additionally, genetic origin can affect fearfulness and feather pecking. The aim of this study was to investigate whether farm conditions and genetic background affect behavior and stress physiology of parent stock flocks of two commonly used commercial hybrids: Dekalb White (n=7) and ISA brown (n=6). Group size per flock was either less than 6,000 (n=5) or more than 6,000 birds (n=8) but with similar stocking densities. We assessed feather damage (n=30/flock, scoring neck, rump and back on a 3-point scale) and fecal corticosterone metabolites (pooled sample of 5 fresh droppings). Further, on six places in the chicken-house a novel object test and a human approach test were conducted. Birds were exposed to a novel object or person for 2 minutes and every 10 seconds the number of birds within 25 cm distance was counted. Flock impression was assessed by a qualitative behavioral assessment (QBA) on a 6-point scale. Data were analyzed by flock average with the GLM procedure, model consisted of line, farm and group size. Dekalb White flocks approached the novel object sooner than ISA flocks (58 vs.119 s ±13 s, $P<0.05$). In large group sizes, QBA scores related to frustration were higher and those related to activity were lower (both $P<0.05$), and differed between farms ($P<0.01$) but not between hybrids. Flocks differed in corticosterone metabolites ($P<0.05$), with considerable variation between flocks and hybrids. No effects were found on feather damage or time of birds to approach the human (3 out of 13 approached). The results of this study did not show a relationship between hybrid and farm conditions on fear and feather pecking, although Dekalb White flocks were more likely to approach the novel object. Group size influenced QBA scores. Possibly, large flocks may be perceived differently by the observer than small flocks. The large variation in fecal corticosterone metabolites between farms can indicate effects of management on the ability to cope with fear and stress. Especially in parent stock flocks, high fearfulness and chronic stress should be avoided to maintain production and welfare. A possible route of improvement can lie in improving human-animal interaction, as the response to a human may indicate fear for the farmer.

Practical animal based measures for assessing the effectiveness of a novel stunning device for rabbits

Rau, J.[1], Lawlis, P.C.[2] and Joynes, K.[2], [1]University of Guelph, Ontario Veterinary College, Guelph Ontario, N1G 2H1, Canada, [2]Ontario Ministry of Agriculture, Food and Rural Affairs, Animal Health and Welfare Branch, 401 Lakeview Drive Unit A, Woodstock Ontario N4T 1W2, Canada; jarau@uoguelph.ca

In 2006, the Province of Ontario eliminated the use of manual blunt trauma for the stunning of rabbits at slaughter. Prior to this deadline, Ontario Ministry of Agriculture and Rural Affairs (OMAFRA) staff together with colleagues at the Ontario Veterinary College and plant personnel developed a novel technique to stun rabbits at slaughter using a non-penetrating captive bolt device (Zephyr). The objective of this project was to develop practical animal based measures for assessing effectiveness of stunning and return-to-sensibility of rabbits at slaughter. The measures developed and used in the project were: (1) absence of a corneal eye-blink reflex; (2) absence of jaw tone. The rabbits observed in this study were New Zealand White fryers, weighing approximately 2.5 kg. All rabbits originated from a single supplier and arrived at the abattoir on the morning of the test day. A total of 560 rabbits were observed. A randomized design was applied to assign rabbits to one of two treatment groups – (1) frontal application of the Zephyr or (2) rear approach application of the Zephyr. A single trained abattoir employee carried out all stunning procedures over the course of one day. OMAFRA Meat Inspection Staff trained this employee in the proper technique for stunning using the non-penetrating captive bolt. Each of the treatment groups were observed at the point of stunning and at 30 seconds post-stunning (bleed out). The conventional method of stunning using the non-penetrating captive bolt at the front of the head was effective at producing insensibility for 100% and 99.96% of the observations taken immediately post-stunning and at 30 seconds post-stunning, respectively. Rear-approach stunning was less effective with only 88.26% of the rabbits stunned at the first attempt and 70.43% of rabbits showing signs of return-to-sensibility at 30 seconds post-stunning. The results of this study suggest that corneal reflex and jaw tone are practical methods for assessing the effectiveness of stunning of rabbits.

Can early handling of suckler beef calves reduce their timidity towards humans?

Probst, J.K.[1,2]*, Hillmann, E.*[2]*, Leiber, F.*[2]*, Kreuzer, M.*[2] *and Spengler Neff, A.*[1], [1]*FiBL, Research Institute of Organic Agriculture, Animal Husbandry Division, Ackerstrasse/Postfach, 5070 Frick, Switzerland,* [2]*ETH Zurich, Institute of Agricultural Sciences, Universitätsstrasse 2/ LFW, 8092 Zurich, Switzerland; johanna.probst@fibl.org*

In herds of suckler beef cows the relationship between humans and animals is an important issue concerning animal handling. Unlike dairy cow herds, suckler beef herds are less used to close contact with humans. Due to low management input animals may be difficult to handle during treatments like medical care, routine handling or at slaughter house. Additionally those attitudes can cause injuries to both stock and handler. At this time it should be possible to get in close contact with the calves and establish a positive relationship. We investigated whether positive tactile handling, performed in the postnatal period, can reduce timidity towards humans in calves. A group of total 27 Limousin-crossbred suckler beef calves was randomly allocated to a handling group (HG/ 7 female, 6 male) and an age- and sex-matched control group (CG/ 8 female, 6 male). HG calves were handled tactile on the 2nd, 3rd and 4th day after birth and furthermore on 3 days during the following 3 weeks. Handlings were conducted twice a day in 10 min sessions repeated after 30 min. The handling (without offering feed) was based on the elements of tactile approaches of TTouch®. Handling treatments took place in the home pen or on pasture and were performed by a person unfamiliar to the cows and calves. At ages of 2, 3, 5, 6, 8 and 9 months each calf was tested with the avoidance distance test (ADT). Furthermore any voluntary approach to humans during the ADT was recorded. All calves were slaughtered when they were 10 months old. Inside the stunning box the head positions of animals were recorded and scaled into: (1) animal tried to move backwards, (2) animal stayed neutral, (3) animal moved forward (no sign of fear). Data were analyzed using generalized linear mixed effects models and Pearson's chi-square test. HG calves approached the test person more often voluntarily than CG calves (t_{25}=4.33, P<0.001). HG calves showed shorter avoidance distances towards the test person than CG calves ($F_{1,25}$=18.98, P<0.001). Avoidance distances were generally greater on pasture than in the barn ($F_{1,129}$=19.39, P<0.001). Before stunning HG calves showed less moving backwards behavior compared to CG calves (χ^2_2=13.9; P<0.01). In conclusion positive handling in the early life of beef suckler calves can reduce their timidity towards humans. This attitude persists over a long period and even under the stressful conditions of the slaughter house.

Assessment of horse welfare in the Netherlands: how to train the assessors to achieve reliable animal measurements?

Neijenhuis, F.[1], De Graaf-Roelfsema, E.[2], Wesselink, H.G.M.[3], Van Reenen, C.G.[1] and Visser, E.K.[1], [1]Wageningen UR, Livestock Research, P.O. Box 65, 8200 AB Lelystad, Netherlands, [2]Utrecht University, Equine Internal Medicine, Yalelaan 114, 3584 CM Utrecht, Netherlands, [3]Independent Veterinary Professional, Beerzerweg 11, 7736 PH Beerze, Netherlands; francesca.neijenhuis@wur.nl

The European Welfare Quality® project has produced protocols for the welfare assessment of farm animals, excluding horses. In the Netherlands protocols for welfare assessment for horses have been developed in line with the WQ® system. The quality of an assessment depends greatly on the quality of the measurements. To ensure the quality of the measurements thorough training of the assessor is a necessity. The purpose of this study was to develop a training program for potential assessors for the welfare assessment of horses that ensures the quality of the measurements. The protocols used are based on the Welfare Quality® system, and include the four principles: feeding, housing, health and behavior. The selected measurements were checked for their quality on the following criteria (1) feasibility, (2) validity and (3) reliability. Two senior veterinarians on horse health participated in the study as trainers. They trained five veterinarian students. The following health measures were included: skin abnormalities, wounds, lameness, hoof condition, breathing, coughing, abnormalities in incisors, mouth corners, gums, back pain, coat condition, nasal and ocular discharge. Intra- and inter reliability of the scores was tested using percentage of agreement, Fisher's and McNemar test, binominal test and the Kappa coefficient test. The training program consisted of an audio visual training and a practical training. The two trainers trained themselves to become a 'silver standard'; inter- and intra-reliability was tested with 30 horses. Thereafter they selected photos to serve as a 'golden standard' for the audio visual training. In the practical training the potential assessors and the trainers scored 90-120 horses independently. The agreement between the two trainers was 65% to 100% and within trainers 69-100% (trainer 1) and 62-100% (trainer 2). All potential assessors scored at least 80% correct in the audio visual training. The practical training lasted 6 days. The reliability between the potential assessor and trainer was good on basis of the binomial test, the Fisher and the McNemar test. Between the assessors all measures showed significant agreement (kappa between 0.303 and 0.992, $P<0.005$). For assessing animal welfare trainers and potential assessors need to be trained thoroughly. Special attention should be paid to ensure that horses used in the training show enough variation in scores and that potential assessors have appropriate skills.

Behavioral ecology of captive species: using behavioural needs to assess and enhance welfare of zoo animals

Koene, P., Wageningen University, Wageningen UR LIvestock Research, Marijke weg 40, 6709 PG, Netherlands; paul.koene@wur.nl

Wild animals are adapted to the environment they evolved in (EEA). In relatively stable environments competition between and within species urges animals to be specialists (food, defense, etc.). In variable environments animals have to be adaptive (generalists). Species with specific environmental adaptations may show specific behavioral needs, difficulty in adapting to a new environment, and suboptimal functioning and fitness. Animals in zoos are perceived as representatives of their wild counterparts. Discrepancy between natural behavioral needs and behavioral possibilities in captivity may cause welfare problems. Aim of the project is to estimate a species' suitability for living in captivity, assess welfare, suggest environmental changes, and find species characteristics that underlie welfare problems in zoo animals. First, the current status of zoo animal welfare assessment is reviewed and the new approach is outlined. Databases of species characteristics are set-up using literature of natural behavior (1) and captive behavior (2). Species characteristics are grouped in eight functional behavioral ecological fitness-related categories related to space, time, metabolic, safety, reproductive, comfort, social and information needs using a model of welfare optimization. Assessments of the strength of behavioral needs in relation to environmental needs are made based on results available from literature. The databases with literature on species level are coupled with databases of behavioral observations (3) and welfare assessments (4) under captive conditions. The represented structure produces best professional judgments, shows discrepancies between environmental responses in different environments and suggests ways for improvement (environmental changes). Using phylogenetic correction, actual welfare problems are related with natural behavior and ecology of different species. Behavioral data from many MSc-projects covering 10 Dutch zoos and 45 species are used. The approach is compared with and incorporates principles, methods and outcomes developed in the Welfare Quality® project in the functional behavioral category approach. Observation and welfare assessment methods are adapted from the farm environment and applied to zoo specific environment, using research on tigers, giraffes and wolves. The newly developed methods of observation and welfare assessment were tested in a quick scan of 25 additional species in Dutch zoos (mammals, birds and reptiles). In conclusion, the comparison of the complete behavioral repertoire of behaviors in natural and captive environments highlights welfare problems, the solution of welfare problems by environmental changes and the species characteristics underlying zoo animal welfare problems.

Behavior, management and welfare of sled dogs in the Netherlands

Koene, P. and Hermsen, D., Wageningen University, Department of Animal Sciences, Marijkeweg 40, 6709 PG, Netherlands; paul.koene@wur.nl

Sled dog racing is worldwide a popular sport. Relatively few people – mushers – are actively involved in sled dog racing in the Netherlands. During racing, dogs run the risk of injury, exhaustion or dehydration. Studies in North America show that sled dog welfare may be harmed. In general there is very limited literature about behaviour and welfare of sled dogs during racing, training and in their normal living situation. The aim of this study was to observe behaviour and assess welfare of sled dogs in their home situation. Our general framework for welfare assessment was a modified Welfare Quality® protocol. First a survey was distributed to sled dog owners to collect information about housing, handling and management procedures. One of the questions concerned vists for observation. The survey resulted in responses of 33 mushers (236 sled dogs) and provided a first overview on Dutch sled dog husbandry. Mushers protect their dogs against dehydration by providing bouillon. Sled dogs are protected against hyperthermia by providing shade during housing and by using a maximum temperature for racing (13.7 degrees Celsius). Sport injuries do occur, but they were not severe compared with data from North american studies. Furthermore general health is taken care of very well, and doping use seems to be absent in The Netherlands. Behaviour was observed at the homes of 15 mushers covering 174 Siberian Huskies. Using scan and behaviour sampling time budgets and social interactions of sled dogs were recorded.The time budgets found showed that sled dogs are very passive and that dogs kept on a field were less passive than in other environments (M-W, $U=103$, $P=0.002$). The frequency of social play interactions per hour differed between housing conditions was analysed using a linear mixed model ($F(4,74) = 4.31$, $P=0.003$). On the field (7.80) and in the garden (8.78) more play interactions ($P=0.047$; $P=0.026$) are recorded than in the kennels (1.57). The frequency of fight interactions per hour differed between housing conditions ($F(4,44) = 2.65$, $P=0.046$). More fights ($P=0.030$) are found in the field (0.59) than in the kennel (0.04). In summary, six sub criteria indicated good welfare (dehydration, hyperthermia, sports injuries, absence of health problems, social behaviours and other behaviour) and 4 indicated suboptimal welfare (proper diet, comfort, movement and pain). In general sled dogs show much passive behaviour. Their activity is much higher and more social (play) is found when the dogs are kept on a field during the day. On the negative side, some fights may occur. Sled dogs show more natural behaviour and their welfare seems to be improved when housed on a field during daytime. However, the survey and welfare assessment are preliminary and more research is needed.

Output vs.design indicators: a review of the benefits and drawbacks for on-farm welfare assessment

Veissier, I.[1], Mounier, L.[1], Dalmau, A.[2], Knierim, U.[3], Winckler, C.[4] and Velarde, A.[2], [1]Inra, Theix, 63122 Saint-Genes-Champanelle, France, [2]Irta, Granja, 17121 Monells, Spain, [3]UniKassel, Nordbahnhofst.1a, 37213 Witzenhausen, Germany, [4]Boku, G-Mendel-St.33, 1180 Vienna, Austria; veissier@clermont.inra.fr

On-farm animal welfare assessment may be based on design indicators on the housing, feeding, ... (e.g. ANI) or output measures (i.e. on animals as Welfare Quality®). The choice between these measures is often dogmatic, some authors considering that design indicators are risk factors whereas output measures better reflect the true welfare of animals. To compare design vs.output measures, we drew a list of properties for a measure to be considered valid for assessing welfare. These properties are derived from analytical methods: selectivity, trueness, reliability, stability over time, fitness for the purpose (including sensitivity), and feasibility in different systems. The properties of measures that may be used to check the 12 welfare criteria defined in Welfare Quality®were analysed. For absence of hunger, injuries and diseases; for the expression of social or negative behavior (e.g. stereotypies) and for the relation to humans, output measures (body condition, clinical and behavioral observations) are selective whereas measures on foods, management, human behavior or attitude have a much lower predictive value. Similarly, the emotional state of animals may be inferred from a qualitative behavioral assessment but hardly from the environment. In these cases, output measures are recommended. Positive behaviour (play, exploration) and difficulties in moving (slipping, falling) are rare so difficult to detect. Output measures such as lying behavior or injuries due to frictions may be not sensitive enough to check comfort around resting in some environments. Here, fitness for the purpose is impaired and output measures should be used in combination with design ones. For absence of thirst, tests on animals detect only strong dehydration thus number and quality of water points are more valid. Design measures are also the only possible for assessing pain due to procedures like dehorning when the observer is not present when these occur. For thermal comfort, design measures (i.e. ambiance) are reliable only if repeated extensively (so feasibility is low); aspect or behavior of animals (e.g. huddling vs.panting in pigs) seem more reliable in a short visit. In conclusion, there is no general rule whether output measures are more or less valid than design ones. This depends on the welfare criterion considered. We recommend to use a mixture of output and design measures (chosen for their validity and feasibility) to assess the overall welfare of animals on farms. This work was supported by Scaw.

Inter- and intra-observer reliability of experienced and inexperienced observers for the Qualitative Behaviour Assessment in dairy cattle

Bokkers, E. and Antonissen, I., Wageningen University, Animal Production Systems, POB 338, 6700AH Wageningen, Netherlands; eddie.bokkers@wur.nl

Qualitative Behaviour Assessment (QBA) is part of the Welfare Quality (WQ) protocol in dairy cattle, although inter- and intra-observer reliability has been scarcely studied. It is stated that QBA assessors should be experienced with cattle. This study evaluated inter- and intra-observer reliability for the QBA of dairy cattle in experienced and inexperienced observers using videos recordings. 8 Observers were trained with videos and on-farm instructions in QBA. After training they did a test which was repeated after half a year. The test contained 16 video clips (30 s/clip) showing a few cattle. After each clip observers had to fill out the QBA scoring form, which consists of 20 terms that represent a human interpretation of emotional expressions of cattle (active, relaxed, fearful, agitated, calm, content, indifferent, frustrated, friendly, bored, playful, positively occupied, lively, inquisitive, irritable, uneasy, sociable, apathetic, happy, distressed). A score per term was given on a scale ranging from 0 mm (absent) to 125 mm (dominant). On the same day as the 2nd test, observers scored another 11 video clips showing herds (4×30 s/clip). In between tests, observers conducted several on-farm QBAs. In addition, 10 inexperienced observers did a QBA on both video sets. Inter-observer reliability was analysed per term and for the 1st PCA factor score of individual observers using Kendall's W. Intra-observer reliability was analysed per term for set 1 (paired T-test, Spearman's rank correlation). Inter-observer reliability per term of experienced observers ranged for set 1 from 0.22 (indifferent) to 0.61 (fearful) for the 1st test and from 0.15 (bored) to 0.64 (playful) for the 2nd test and for set 2 from 0.38 (relaxed) to 0.87 (playful). Scores of inexperienced observers ranged for set 1 from 0.34 (active) to 0.68 (friendly) and for set 2 from 0.23 (active) to 0.89 (distressed). Inter-observer reliability of experienced observers based on the factor score was 0.35 in test 1 and 0.45 in test 2 for set 1 and 0.78 for set 2. For inexperienced observers it was 0.64 and 0.73 for set 1 and 2. Intra-observer correlations (r_s) ranged from -0.34 to 0.97 and paired differences from 0.1 ($P=0.99$) to -35.9 ($P<0.01$), whereas high correlations were not necessarily associated with low paired differences. Inter-observer reliability was low to moderate for individual QBA terms and low to good for the factor score. Experience did not improve inter- and intra-observer reliability. Type of video clips influenced reliability. Intra-observer reliability varied largely per term and per observer. Overall, QBA does not seem to be a reliable tool within the WQ protocol for dairy cattle.

Development of a welfare assessment protocol for caged gamebirds during the laying season

Matheson, S.M., Donbavand, J., Sandilands, V., Pennycott, T. and Turner, S.P., SAC, West Mains Road, Edinburgh, Scotland, EH9 3JG, United Kingdom; Stephanie.Matheson@sac.ac.uk

Each year the UK rears around 20-30 million pheasants and 3-6 million red-legged partridges for shooting, however, the incidence and severity of the challenges to gamebird welfare during captivity are poorly understood. Of particular concern is the use of barren cages for breeding gamebirds, which share some of the characteristics of battery cage systems used for commercial egg production in poultry. The Farm Animal Welfare Council and the gamebird industry itself have voiced concerns that such systems are incompatible with their ethical values, suggesting that the welfare of gamebirds in cages justifies rigorous assessment. The focus must be on identifying the biological needs of the birds involved in order to find the appropriate design criteria for caged laying systems, one which optimises both welfare and animal production. The needs of the industry must be matched with the needs of the laying birds (e.g. industry wants ease of egg collection but this must be tempered with the hens' need for a secluded area in which to lay). There are several unique challenges associated with caged gamebirds. In the UK, the use of cages for breeding is currently unregulated, falling as it does between the domestic and wild animal legislation. It is, in essence, an intensive breeding and laying system in an outdoor situation, with all the disadvantages of an extensive system (such as varied weather conditions). In addition, gamebirds are semi-wild species, thus the nature of gamebirds, which is essentially between domestic and truly wild animals, adds an extra dimension to welfare assessment. Therefore, caged environments must take into account the breeding ecology of the species in question and, importantly, allow the birds to display their full repertoire of biologically significant behaviours. There is some justification for drawing upon findings from the poultry welfare literature, however, these cannot be implemented without refinement and validation with regards to the specific needs of gamebird species. With these considerations in mind, we used the Farm Animal Welfare Council's concept of the Five Freedoms as a basis for assessment, resulting in a welfare protocol designed for use on-farm. This protocol was then further refined within the context of a commercial breeding farm. With these refined welfare indicators in mind, the next challenge is to design a system that maximises the welfare of captive breeding gamebirds while maintaining an economically efficient and sustainable enterprise.

Thermal nociception as a tool to investigate NSAID analgesia in a model of inflammatory pain in broilers

Caplen, G.[1], Hothersall, B.[1], Sandilands, V.[2], Mckeegan, D.[3], Baker, L.[2], Murrell, J.[1] and Waterman-Pearson, A.[1], [1]University of Bristol, Langford, BS40 5DU, United Kingdom, [2]SAC Auchincruive, Ayr, KA6 5HW, United Kingdom, [3]University of Glasgow, Glasgow, G61 1QH, United Kingdom; gina.caplen@bristol.ac.uk

A better understanding of pain associated with lameness in broilers requires development of appropriate analgesic protocols for pain management. This study utilised a thermal threshold testing device (TopCat Metrology) to investigate how thermal nociception differs between birds artificially induced to become lame with those that remain sound, and examined the effect of NSAID treatment. Ross broilers (GS1, n=48 total) were subjected to both a 'treatment' (saline control, 3 or 5 mg/kg meloxicam, 15 or 25 mg/kg carprofen: administered subcutaneously), and an 'experimental procedure' prior to thermal threshold testing. The latter comprised either 'lameness induction' (injection of 0.4 ml Freund's adjuvant into the left hock), or sham handling. Threshold testing was performed via application of a ramped thermal stimulus to skin covering the left metatarsal, until a behavioural response was observed or a pre-specified cut-out temperature reached. Statistical analysis was conducted using Independent-Samples T-Test (SPSS 17.0); df for all tests = 14. All induced hocks were visibly swollen. Skin temperature (±SD) of lame-induced birds was significantly higher than sham-handled birds for saline (37.60±0.80 °C/36.54±0.45 °C; t=-3.25, P=0.006) and meloxicam (36.95±1.02 °C/35.40±0.67 °C; t=-3.702, P=0.002), consistent with inflammation. Saline-treated birds with induced lameness had significantly lower excursion thresholds than their sham-handled counterparts (5.15±1.09 °C/7.15±2.09 °C; t=-2.40, P=0.036), suggesting that the swelling and lameness were associated with hyperalgesia. Analgesic-treated lame-induced birds had excursion thresholds comparable with saline-treated sham-handled birds (meloxicam: 7.39±1.71 °C; t=-0.252, P=0.805; carprofen: 7.65±1.41 °C; t=-0.563, P=0.582), suggesting that analgesic treatment restores heat sensitivity to 'normal'. Sham-handled birds exhibited comparable excursion thresholds regardless of treatment; excursion thresholds in analgesic-treated lame-induced birds significantly increased above that observed with saline in lame birds (meloxicam: t=-3.117, P=0.008; carprofen: t=-3.954, P=0.001). This may be expected if the analgesic compounds specifically target the site of inflammation in the lame-induced cohort, as opposed to a 'less effective' wider systemic distribution within the sham handled cohort. Results indicate both analgesics have potential for providing pain relief to birds with induced hock-joint inflammation; both appearing equally effective at inducing hypoalgesia at the doses administered.

Improving the practicality of measuring motivation for food in ruminants

Doughty, A.K.[1,2], Ferguson, D.M.[1], Matthews, L.R.[3] and Hinch, G.N.[2], [1]CSIRO, Armidale, NSW 2350, Australia, [2]University of New England, Armidale, NSW 2350, Australia, [3]AgResearch, Hamilton, 3214, New Zealand; Amanda.Doughty@csiro.au

The measurement of strength of motivation is widely used to assess the resources that an animal values by providing an insight into how much work an animal is willing to do to obtain that resource. Animals are trained to perform an operant behaviour to access a resource and are then asked to 'pay a price' to maintain this access. A proven protocol for testing food motivation in ruminants allows unrestricted access to food over 23 hours, meaning that only one animal can be tested at each access price in each set of apparatus per day. When testing is added to the necessary habituation and training required, motivation experiments can take upwards of four months to complete. As the equipment is generally too expensive to replicate it would be useful if several animals could be tested per day. Therefore, the aim of this investigation was to determine whether shorter periods of testing could differentiate food motivation between animals exposed to different levels of food restriction. Eight sheep were trained in a 50 m U-shaped laneway to access a double-sided feeder and gained a 5 g food reward with each access event. Sheep were then tested to see how many times in a 23 h period they would walk a specific distance for the reward. The distance (cost) that the sheep walked was increased progressively on a log scale (1.5-105.5 m). Sheep were randomly allocated to one of two treatments (14 h restriction and an unrestricted control). Data were transformed to correct for heteroskedasticity and analysed using linear regression. The demand function was calculated as $\text{Ln}(Q) = \text{In}(L) + b[\text{In}(P)] - a(P)$ and a benchmark of 70% (percentage of variance accounted for) was set as a suitable figure to show differentiation between motivation levels. The results showed that after nine hours in the test facility, the control sheep had completed 69.1% (S.E.M ± 2.7) of the work that they would do over the available 23 h period while the restricted sheep completed 87.5% (S.E.M ± 2.1) of the work. Investigating a shorter time frame showed that at 3 h 53.0% (S.E.M ± 2.2) of the work had been completed by the control sheep compared with 74.4% (S.E.M ± 2.4) by the restricted sheep. These data therefore suggest that it may be possible to substantially reduce the length of time necessary to differentiate motivation levels for food in a ruminant. The capacity to reduce the duration of a test would increase the numbers of animals tested daily and therefore reduce the total duration of the experiment. However, additional studies are necessary to better understand how these testing procedures may be further modified to make this experimental methodology even more practical.

Estimating deep body temperature in animal welfare studies; non-invasive alternatives?

Mitchell, M.A.[1], Farish, M.[1], Kettlewell, P.J.[1] and Villarroel-Robinson, M.[2], [1]SAC, Sustainable Livestock Science, Sir Stephen Watson Building, Bush Estate, Penicuik, EH26 0PH, United Kingdom, [2]Universidad Politecnica de Madrid, Escuela Técnica Superior de Ingenieros Agronomos, Avenida Complutense s/n, 28040 Madrid, Spain; malcolm.mitchell@sac.ac.uk

Deep body temperature (DBT) is a widely employed clinical indicator of infection, physiological and psychological status and responses to environmental challenges and stress. DBT has been widely employed as an indicator of stress in studies relating to the welfare of animals during a number of handling, transport and slaughter procedures. Most routine methods of measuring DBT tend to be minimally invasive and provide estimates of true core temperature with varying degrees of accuracy e.g. rectal or vaginal temperature, ear temperature or even skin temperature. These traditional methods require animal handling and restraint which are undesirable in many experimental contexts. Invasive methods e.g. surgically implanted devices provide more accurate measures of core temperature but are more complex, demanding and costly than the non-invasive alternatives. The present study has evaluated two minimally invasive approaches, not requiring restraint, by comparison with implanted telemetry devices. Injected or implanted (sub-cutaneous) RFID chips (Identichip-Biothermo) that contain a temperature sensor and skin temperature (ST) measured by non-contact infra-red thermometry (Raytek MX4) have been employed to estimate body temperature in pigs under a range of controlled thermal conditions at temperatures between -10 and +35 °C during exposures of up to 8 hours duration in controlled climate chambers. Implanted chip temperatures exhibited poor correlation with core temperatures across the full environmental temperature range but during exposure to elevated thermal loads DBT was predicted accurately by deep implanted (3 cm) chips ($y = 0.898x + 3.42$; $R^2=0.6$, $P<0.01$). Similarly surface temperatures measured at 8 different anatomical sites were poorly correlated with DBT ($y = 2.63x + 72.7$; mean $R^2=0.127$) but were significantly correlated with air temperature (mean $R^2=0.91$; $P<0.005$) whereas when only the relationships at air temperatures of 29 °C or greater were considered significant predictive correlations (mean $R^2=0.56$; $P<0.05$) between ST and DBT were found. It is concluded that these minimally or non-invasive methods for DBT estimation are of limited use during cold exposure or at thermo-neutral air temperatures but offer useful and valuable alternatives for characterization of thermoregulatory challenge at elevated environmental heat loads and may be usefully applied in both laboratory trials and under commercial animal production conditions in the assessment of thermal stress and animal welfare.

Applying Welfare Quality® strategy to interpret and aggregate welfare measures for farmed fur animals

Gaborit, M., Veissier, I. and Botreau, R., INRA, UR1213 Herbivores, F-63122 Saint-Genès-Champanelle, France; raphaelle.botreau@clermont.inra.fr

The WelFur project aims to develop on-farm welfare assessment protocols for farmed mink and foxes. These protocols are based on the Welfare Quality® (WQ®) system developed for cattle, pigs and poultry. This system is based on a hierarchical structure, where 4 principles of welfare (Good feeding, Good housing, Good health, Appropriate behaviour), subdivided into 12 criteria, are constructed and then aggregated to obtain an overall evaluation at farm level. For mink and foxes, partners of WelFur developed measures to check the compliance of fur farms with each of the 12 criteria. We apply the multicriteria decision-aiding methods developed in WQ® to model the interpretation of measures in terms of welfare and aggregate the results, and thus calculate the 12 criterion-scores. As in WQ®, the evaluation model is parameterized using experts' opinion. Two expert panels, one for mink and one for foxes, were established, with several animal scientists chosen for their knowledge on the two species in terms of physiology & behaviour and farming practices. Even if the general construction procedure is the same as in WQ®, several specificities emerged. First, contrary to species present in WQ®, here the whole production cycle, from breeding to slaughter of grown juveniles, occurs on the same farm. Thus to have an overall view of the whole fur farm, three periods of the production cycle (pelting to mating / mating to weaning / weaning to pelting) should be assessed. Depending on the period, the number and types of animals (adult males, adult females, under-weaning and growing juveniles), and the resources used differ. This has a direct impact on the construction of the criteria, e.g. at criterion level we have to integrate the data collected at several periods and this requires specific arrangements for the calculation of scores. Second, two different species of foxes and their hybrids are to be considered. They differ in behaviour and needs, and thus the interpretation of a given measure can vary. For the calculation of criterion-scores, when appropriate, we have to adjust the construction and run separate consultations of the experts. Third, the assessment systems developed in WelFur should be at least applicable on all the production systems present in Europe, including variability in regulations (e.g. size of cages) and climatic conditions. Hence, when interpreting the measures, the experts need to take into consideration the whole range of existing variability. This work shows that the rationale followed in WQ® can be extended to develop assessment systems for other species once taking into account the specificities of these species or the production systems.

Impact of maternal equine appeasing pheromone (EAP) during a short term transport in saddle horses

Cozzi, A., Lafont-Lecuelle, C., Articlaux, F., Monneret, P., Bienboire-Frosini, C., Bougrat, L. and Pageat, P., IRSEA Research Institute Semiochemistry and Applied Ethology, Pets and Sport Animals Department, Le Rieu Neuf, 84490, Saint Saturnin Les Apt, France; a.cozzi@irsea.info

Controlled studies in species like cat, dog, chicken, pig and horse showed the interest using semiochemicals in order to manage the process of adaptation during a stressful situation. Studies performed on horses traveling showed the modification of behavioral and physiological indicators linked to stress. The present study was designed to investigate the impact of the EAP gel during a short term road transport for horses that regularly travel short distances. 28 horses were included in the study and the experimental design was as follows: double blinded, randomized, two parallel groups. 17 horses and 11 ponies, from different breed, between 2 and 20 years old, were divided in two homogeneous groups for sex, age and transport experience. The gel was applied into the nostril 20/30 minutes before the transport. Horses were transported for 1 h on a defined 60-km route; any injured was detected after transportation. Horses traveled alone in the same truck with the same driver. Two transports per day have been done between 8.30-13.30 am. A cleaning protocol of the truck after each transport have been done. The level of stress during and after transport was described with the measure of Heart Rate (Polar Horse S810®), salivary cortisol, frequency of neighing during the trip and water loss in the dung recovered in the truck. The placebo and EAP groups were compared for all parameters during and at the end of the transport using the two-sample Student t test (α=5%). We found significant differences between the two groups in favor of EAP for the comparison of Heart Rate (mean ± standard deviation: (EAP: 56.36±6.70 bpm; placebo: 64.48±8.62 bpm; ddl=26; t=2.79; P=0.0098) and water loss in the dung (EAP: 71.84±5.05%; placebo: 76.48±4.70%; ddl=26; t=2.52; P=0.0183); we found no significant differences between the two groups for salivary cortisol (EAP: 0.42±0.16 µg/dl; placebo: 0.48±0.21 µg/dl; ddl=26; t=0.87; P=0.3913) and for the frequency of neighing during travel (EAP: 8.64±9.30; placebo: 8.29±10.02; ddl=26; t=-0.10; P=0.9229). Results show the interest to use the Equine Appeasing Pheromone in order to manage dehydration and changes in cardiac parameters during and immediately after a stressful event like transport. Transporting horses is not insignificant, even on those that are reported to be confident with the event; the present study highlighted the interest to use the semiochemical approach in horses during short term transport to facilitate the adaptation process naturally, without using forbidden substances.

Objective assessment of pain in dairy cattle with clinical mastitis

Fitzpatrick, C.E.[1], *Chapinal, N.*[1,2], *Petersson-Wolfe, C.S.*[3] *and Leslie, K.E.*[1], [1]*University of Guelph, 50 Stone Road East, Guelph, ON, N1G 2W1, Canada,* [2]*University of British Columbia, 2329 West Mall, Vancouver, B.C., V6T 1Z4, Canada,* [3]*Virginia Polytechnic Institute and State University, 118 N. Main St., Blacksburg, VA, 24061, USA; cfitzpat@uoguelph.ca*

Clinical mastitis is a prevalent problem in the dairy industry, and has detrimental effects on the animal's profitability, as well as negative impacts on cow welfare. As with many disease conditions in animals, it is inferred that mastitis causes significant discomfort and pain. There are many studies that present extremely useful information about discomfort with mastitis, however, published documentation quantifying pain with mastitis is not abundant. The current study was conducted to objectively assess pain in cases of experimentally-induced clinical mastitis, to better understand the effects of clinical mastitis on dairy cattle welfare. In August 2010, 24 dairy cows from the dairy research centers of the University of Guelph were enrolled in a LPS challenge study. Each animal was challenged in one rear mammary quarter by intramammary infusion with 25 µg of *E. coli* LPS. Subsequently, a subcutaneous injection of either a placebo (n=12) or NSAID treatment (meloxicam) (n=12) was randomly allocated and administered using, yet to be identified, double-blind methods (Treatments A and B). The animals were monitored for two days prior to, and two days following, the intramammary challenge. Several behavioural, physiological and performance parameters were monitored throughout the study period. The data was analyzed using the mixed procedure in SAS. During the first six hours after inoculation and treatment, cows ruminated 14.6±2.1 min/2 h interval ($P<0.001$) less compared to the same baseline time period prior to challenge. Overall, multiparous cows were found to ruminate 6.1±1.6 min/2 h ($P=0.001$) more than primiparous cows. There was no difference in rumination between treatment groups. Using a pain pressure algometer, the difference between the pressures applied to the control quarter was compared with the pressure on the challenge quarter. There was an effect at hour six after inoculation and treatment as compared to the baseline readings. For Treatment A animals, more pressure could be applied on their challenge quarter than their control quarter (1.9±0.9 lbs) ($P=0.0445$). Treatment B animals registered more pressure applied to the control quarter than the challenge quarter (2.5±0.9 lbs.) ($P<0.01$). These results indicate the potential for using continuous measurement of rumination and pain pressure sensitivity for objective assessment of pain due to illness in cases of clinical mastitis. These methods can be used to identify discomfort in an animal, which will hopefully allow for pain intervention, and an increased welfare for the animal.

Development of precision livestock farming solutions for animal health and welfare

Roulston, N.[1,2], Romanini, C.E.B.[1,3], Bahr, C.[1], Birk, U.[1], Demmers, T.[1], Eterradossi, N.[1], Garain, P.[1], Guarino, M. [1], Halachmi, I.[1], Hartung, J.[1], Lokhorst, K.[1], Vranken, E.[1] and Berckmans, D.[1], [1]BIOBUSINESS Project of the European Union, Training in Research, Product Development, Marketing and Sales in Bio-Business., Marie Curie Initial Training Networks: www.bio-business.eu., Belgium, [2]Fellow PETERSIME nv, Centrumstraat 125, B-9870 Zulte (Olsene), Belgium, [3]Fellow M3-BIORES – Katholieke Universiteit Leuven, Kasteelpark Arenberg 30, B-3001 Leuven, Belgium; nroulsto@uoguelph.ca

Animal production has developed into a highly mechanised and competitive industry. While automating labour intensive tasks such as feeding and climate control have been essential for large-scale production, other animal husbandry practices such animal care have been left behind. In systems which house thousands of animals in restricted space, with high animal turnover and low per animal profit margins, the daily care of animals is expensive, time-consuming, and impractical. Farmers visually scoring individual animals for health and welfare no longer represent the reality of modern livestock production. This scenario has revealed an increase of livestock health and welfare related problems. Combining new technology with animal biology and behaviour can help solve the welfare problems facing modern livestock production. Those educated in animal science, however, are not always aware of the possibilities of modern technology and those developing new technology may not familiar with the world of biology. The main objective of the four-year BioBusiness project is to team up biology-educated people (animal scientists, veterinarians, and biologists) with technology-driven people (bio-engineers and computer engineers) to develop real-time and automated welfare solutions. A consortium of ten partners from research institutions, universities, and industry, working in the field of Precision Livestock Farming (PLF), guides the BioBusiness project. A total of eleven early stage researchers have been recruited as Marie-Curie fellows to develop high-tech PLF products for three projects: (1) automatic improvement of incubation conditions for chicken post-hatch performance and welfare; (2) automatic lameness detection in dairy cows using predictive mathematical models; and (3) automatic monitoring and control system for aggression in group-housed pigs. In order to bring welfare technology to market, fellows are trained in the entire production chain from applied animal research, technology, and product and process development to product and market introduction. The research outcome will be innovative product concepts and corresponding business models of PLF systems for the improvement of livestock health and welfare.

Interest of MHUSA (Mother Hen Uropygial Secretion Analogue) in the management of stress in caged laying hens

Pageat, P., Lecuelle, C., Seurin, F., Bougrat, L. and Saffray, D., IRSEA (Institut de Recherche en Sémiochimie et Ethologie Appliquée), Livestock Species, IRSEA Research Centre – Le Rieu Neuf, 84490 Saint Saturnin les Apt, France; p.pageat@irsea.info

Caged laying hens have been reported to face various stressing events between which the adaptation period, just after the young hens arrived in the production facility. Genetic selection as well as the evolution of the cages used by farmers, have provided some improvement, but failed in erasing the main consequences of this stress. During the last 15 years, numerous studies have emphasized the interest of the use of synthetic pheromones, in the management of stress in mammalian species. More recently, such approach has been shown to be promising in poultry with the description of MHUS (Mother Hen Uropygial Secretion), a maternal uropygial secretion released by mother hens having chicks. The synthetic analogue of this secretion (MHUSA for MHUS Analogue), has been shown to be effective in preventing and managing stress and its detrimental consequences on performances, in broilers. The purpose of this study is to assess the possible effects of MHUSA for the management of stress in laying hens. Two buildings, B1 and B4, welcoming 16,500 and 11,000 hens and located in the same production facility, have been studied. The study has lasted for 2 production periods of 53 weeks, meaning 2 different flocks of hens for each building, during their whole production. During the first year, building B1 received the product when B4 was not treated, and we reversed the situation for the second year. During the first year, the birds belonged to the Isa Brown strain and, for the second year, to HyLine Brown. We studied three parameters, the mortality, the laying ratio and the weight of the eggs. The experimental unit was the week of production. For the statistical analysis we matched the weeks, taking thus in account the season and the strain. We used paired Student-t Test. During the first 28 weeks of production, MHUSA had a significant positive effect on the mortality, assessed by the rate of surviving hens (99.78+/-0.15 for MHUSA, compared to 99.62+/-0.33 for the reference group, P=0.0008), the laying ratio (79.44+/-30.1291 for MHUSA compared to 75.20+/-33.55 for the reference group, P=0.01) and the weight of the eggs (52.02 for MHUSA compared to 51.63 for the reference group, P=0.0195). This effect is also observed, taking in account the whole production period (53 weeks). MHUSA appears to be effective in controlling the stress faced by the laying hens and especially during the first weeks of production, identified as the most stressing period in their life. Such a poorly invasive and polluting treatment looks very promising for the birds living in intensive farming conditions.

Making pig welfare improvements possible health and welfare planning on austrian organic pig farms

Leeb, C., Bernardi, F. and Winckler, C., University of Natural ressources and Life sciences, Department of sustainable agricultural systems, Gregor mendelstrasse 33, A 1180 Vienna, Austria; christine.leeb@boku.ac.at

Improvement of pig welfare remains a huge challenge. After diverse experiences with the UK farm management tool 'Herd Health Plan' including positive aspects and problems, 'BEP BioSchwein' was introduced to improve health and welfare whilst reducing medicine use on Austrian organic pig farms. A farm specific planning tool was developed, integrating quantifiable data, considering farm economy and a farmer orientated process of setting goals and measures to achieve those. At the same time effectiveness, acceptance and economic impact were evaluated. On 60 farms (20 breeders, 20 breeding to finishing, 20 finishers) of relevant size (median 57 sows (8-500); median 175 finishers (65-800) management, feeding and housing were assessed. A representative sample of pigs was evaluated by behavioural observation and clinical parameters based on an adapted WQ®protocol. Animal based assessments and treatment incidences were benchmarked. Based on this, goals (median 2.8 goals/farm) and measures (median 1.2/goal) to achieve those were decided by the farmer. Goals were categorised as health issues (107×), improving productivity (27×), optimising feeding (23×) and housing (12×). Another visit was carried out and after one year a final assessment took place (3.9 visits/farm). Degree of implementation of agreed measures was evaluated, animals assessed to evaluate effectiveness of the tool and the health plan was updated. Achievement of goals was evaluated using predefined parameters for each goal. Degree of implementation of measures was significantly correlated with the success regarding individual goals (Chi^2-Test, $p=<0.0001$, $n=165$). Overall some important parameters improved significantly (Wilcoxon-Test): poor body condition in pregnant (13.8% vs. 9%) and lactating sows (15.5% vs. 8.3%), bursitis in pregnant sows (60.7% vs. 53.7%) and respiratory symptoms in weaners (53.9% vs. 27.6%). Prevalence of pale suckling piglets increased (1.2% vs. 2.4%), whilst all other parameters remained unchanged. Farms aiming at reducing tail biting had a significantly higher reduction of short tails than other farms (-31.2% vs. <-3.4%). Apart from increasing treatment of lameness in sows (3.3% vs. 6.7%) and less treatments of streptococcal disease in suckling piglets (0.7% vs. 0.1%), medicine use remained unchanged. Farmers highly agreed with the tool, specifically with the strategic approach. 'BEP' can be regarded as an innovative and practical advisory and management tool, which can be applied by various stakeholder such as Animal Health Service, (organic) farming associations or advisory bodies.

The acute phase protein, haptoglobine: a potential parameter in welfare assessment?

Ott, S.[1], Moons, C.P.H.[2], Bahr, C.[3], De Backer, K.[1], Berckmans, D.[3], Ödberg, F.O.[2] and Niewold, T.A.[1], [1]K.U.Leuven, Livestock-Nutrition-Quality, Kasteelpark Arenberg 30, 3001 Heverlee, Belgium, [2]Ghent University, Animal Nutrition, Genetics and Ethology, Heidestraat 19, 9000 Gent, Belgium, [3]K.U.Leuven, M3-Biores, Kasteelpark Arenberg 30, 3001 Heverlee, Belgium; sanne.ott@biw.kuleuven.be

Physiological parameters are important measures in animal welfare assessment. To assess the amount of stress an animal experiences, stress hormones like cortisol are frequently used. However, measuring cortisol has major disadvantages due to its rapid reactivity and decline and many influencing factors. Other potential alternative markers are acute phase proteins, since stress is known to affect the immune system. A pilot study was conducted to investigate the response of the acute phase protein, plasma haptoglobine (HP), in pigs subjected to a stressor (food deprivation) and to examine the correlation between HP levels and average daily growth (ADG). Forty grower pigs (25.1±4.4 kg, mean ± SD) (sex and former pen mates balanced), were allocated to 4 conventional pens, 2 treatment (T) and 2 control (C) groups (10 pigs per pen). After 10 days of adaptation the experiment started and ran for 3 weeks. In the 2nd week, T groups were repeatedly subjected to an 8-hour food deprivation (day 1, 3, 5 and 7 of week 2), C groups had normal, unrestricted, access to food. Pigs were weighed twice a week and blood was collected once a week (every 5th day). Mean levels of plasma HP of C and T groups showed large variation between individuals (C groups, week 2: 1.84±3.11 mg/ml; T groups, week 2: 1.40±1.16 mg/ml). No significant differences (Kruskal-Wallis test) in HP levels or growth were found between the C and T groups or between the different weeks within the T groups. Significant negative weak to moderate correlations were found between ADG and HP levels (HP week 1 and ADG week 1: $r_{s=}$-0.47, P=0.005; HP week 2 and ADG total; r_s= -0.60, P=0.015; HP week 3 and ADG total: $r_{s=}$-0.43, P=0.025; average HP total and ADG total: r_s= -0.41, P=0.017). Large variations in HP levels between individuals were shown and no effect of treatment on HP levels or growth was found. Possibly, food deprivation had no apparent stress eliciting effect. Despite these results, interesting correlations between the level of HP and ADG were found, corroborating the inverse relationship between the acute phase response and growth. To further investigate the relation of the acute phase response and stress a successive experiment will be conducted in which we apply a stronger stressor (mixing pigs) and combine the physiological data with behavior.

Developing a new acceleration sensor for measuring the lying time of dairy cows

Tamminen, P., Hakojärvi, M., Häggman, J., Tiusanen, J. and Pastell, M., University of Helsinki, Department of Agricultural Sciences, P.O. Box 28, 00014 University of Helsinki, Finland; petro.tamminen@helsinki.fi

Daily lying time of dairy cows can be used to assess their welfare. There are many factors that are shown to be related to lying time, such as milk yield and leg and claw diseases. Leg-mounted sensors have given good results in previous studies, but the sensors have been relatively large-sized and the battery lifetime has been very limited. Some sensors record the data in internal memory. We aimed to develop a wireless 3D-acceleration sensor that is capable of measuring the daily lying time and animal activity in large herds for long time periods both inside the stall and out in the pasture. Our sensor registers acceleration and inclination in three axes. The measuring range can be selected between 1.5 ... 6 g. Measuring frequency can be freely set, but the maximum is 60 HZ. The on-board nRF9E5 microcontroller communicates internally over SPI-bus, and the measuring data is sent over 869 MHz radio channel. With 2,600 mAh battery, 60 Hz measuring rate and radio power set to maximum the sensor can send data about two weeks and the communication range is about 1 km in pasture and 100 m in stall conditions. Reducing the measuring frequency increases the battery lifetime up to two months. In order to further extend the battery-life there is an option to put the sensor in sleep-mode when no movements are detected. The study was made in the School of agriculture and forestry barn, Seinäjoki university of applied sciences. We recorded the 10 Hz acceleration data from three cows. Recording lasted one week for every cow. Simultaneously we filmed the behavior of cows to validate the data. The analysis was based on the median-filtered acceleration data that was compared to calibration values. A simple algorithm divided the observations into two classes. Lying bouts that were shorter than one second were interpret as false. The measured lying times for cows during one week were accurate. The average total lying time was 4,260 minutes (71 hours) during the one-week experiment. The average error in measurement was +-11 minutes. Average lying bout duration was 44 minutes. The average error in measurement was <+-1 minutes. That is to say, we were able to measure the lying time with over 99% accuracy. The lying time of dairy cows can be measured very accurately wit leg-mounted acceleration sensor. We have developed a wireless acceleration sensor that offers many improvements when comparing to previous devices. After this successful study we have made some further improvements to our measuring system and we are currently measuring the lying time and activity of 43 dairy cows in claw and foot health project.

Technical description and prelimanry results of subcutaneous ecg, activity and temperature logger for dairy cattle

Riistama, J.[1], Vuorela, T.[1] and Saarijärvi, K.[2], [1]Tampere University of Technology, Korkeakoulunkatu 3, 33101 Tampere, Finland, [2]MTT AgriFood Research Finland, Halolantie 31A, 71750 Maaninka, Finland; kirsi.saarijarvi@mtt.fi

There is a need to develop new devices for long term measuring of physiological parameters related to animals' wellbeing. Based on results from several species heart rate variability (HRV) is a promising parameter indicating stress. However, HRV monitors that are placed on the skin do not function well with animals, which have its effect on the measured heart activity data. The authors have developed and tested a continuously working, implantable measurement device able to measure HRV, activity and subcutaneous temperature from a dairy cow. The developed implant is low price, can be modulated and is fairly easy to apply. It was equipped with surgeon steel electrodes and the ECG signal was measured with an instrumentation amplifier. The sampling rate for the ECG was set to 341.3 Hz. The microcontroller used in the device was MSP430F1611 by Texas Instruments. Activity was measured with a 3D-accelerometer. The temperature was monitored with a digital thermometer. The data was written onto an SD memory card. The implant was coated with a hybrid type coating of Parylene-C and medical grade epoxy and it was ethylene oxide sterilized at 42 °C. The estimated operation time was around 30 days. Six *in vivo* -experiments were conducted with the prototype. The device was placed on the left-hand side of the cow by a small operation. The skin and subcutaneous tissue were separated underneath the incision wound to make a pocket under the skin where the device was placed into. The wound was closed with 3-6 sutures. When the *in vivo* -period was over and the data were transferred from the implant to the computer, it was observed that the data was recorded only for one week after the start of the measurement. However, the problem can be corrected by using longer data files. The obtained data was of good quality: out of six hours measuring period 5 h 20 min data was good without any corrections. The quality of data enables straightforward analysis to be used in the signal processing. To make the HRV analysis, the repetitive peaks of the ECG have to be identified. The strongest peak in the cattle ECG is the S-peak. HR can be calculated with a fairly simple 5-point derivative peak detection algorithm accompanied with peak interpolation. The correlation between the accelerometer signals and variations in the ECG signal was calculated and a correlation of 0.39 was found. The implant provides a possibility to separate between physical and psychological stress. It also could provide a possible tool for continuous observing the wellbeing of animals that are under animal welfare surveillance.

Use of eye white and heart rate as dynamic indicators of welfare in Holstein dairy cattle

Tucker, A.L.[1], Zouaoui, B.[2], Devries, T.J.[1] and Bergeron, R.[1], [1]University of Guelph, Department of Animal & Poultry Science, Guelph, ON, N1G 2W1, Canada, [2]National Agronomic Institute of Tunisia, Department of Fish and Animal and Agri-Food Technologies, Tunis, Tunisia; atucker@uoguelph.ca

The percentage of visible eye white (% EW) has been recently identified as a novel and dynamic indicator of emotion in Norwegian Red dairy cattle. Fear and frustration have been associated with opening of the eye (more visible eye white), while positive experiences, such as feeding and calf-cow reunions, have been associated with a reduction in eye white. Heart rate (HR) has also been used extensively as an indicator of welfare in cattle with negative experiences tending to increase cardiac output. The aim of this study was to examine the association between % EW and HR under both a positive rewarding stimulus and a negative fearful stimulus in the most common breed of dairy cattle in North America, the Holstein. Over a 3-d period, 12 multiparous Holstein cows were subjected to both a positive stimulus (provision of food and water) and negative stimulus (startling with umbrella). Heart rate monitors were used to record heart rate at 5-sec intervals for 5 min before and after each treatment while 1 fixed video camera collected behavioural data continuously. A second hand-held video camera (held at a distance of 2 m lateral to the cow) collected eye white data. Still eye white photos were selected for analysis when the cow was facing perpendicular to the hand-held camera and when the cow was neither distracted nor performing any other behaviours (e.g. grooming). Only 1 photo was collected from each 30-sec period and was digitally analyzed. For each stimulus, a general linear mixed model was used to assess the association between % EW and HR; cow and day were included as random effects in the model. Body condition score, weight, parity, stage of lactation and time of day were included as covariates. Heart rate was positively associated with % EW under both the positive (P=0.003) and negative stimulus (P=0.004). Prior to feeding, mean HR and % EW were 94.3±8.6 bpm and 30.2±1.8%, respectively; these decreased to 91.9±5.9 bpm and 27.9±1.2% after feeding (P=0.013). Before opening the umbrella, mean HR and % EW were 74.3±2.2 bpm and 26.5±0.87%, respectively; these increased to 76.5±4.9 and 30.7±1.8% (P=0.099) after opening the umbrella. Closure of the umbrella resulted in a decrease in HR and % EW (P=0.012) to 75.8±2.5 bpm and 26.3±1.3%. These results indicate that EW responses in Holstein cattle are consistent with those reported for Norwegian Red cattle and that % EW is strongly associated with HR under both positive and negative conditions. We suggest that % EW be used as a non-invasive measure for assessing affective state in Holstein dairy cattle.

Thermographic techniques to assess welfare during teeth grinding in piglets

Redaelli, V.[1], Luzi, F.[1], Verga, M.[1] and Farish, M.[2], [1]Univ. of Milano, DSA, via G. Celoria 10, 20133 Milano, Italy, [2]SAC, Bush Estate, EH26 0PH Penicuik, United Kingdom; veronica.redaelli@unimi.it

Teeth resection of piglets within 24 hours of birth is a practice commonly applied in commercial farms to avoid facial injuries of littermates during establishment of teat orders and long term damage to the sow's udder. Opinion about resection and techniques is not unanimous among researchers and the debate is very heated. A preliminary study was implemented to ascertain whether novel thermographic techniques (IRT) could be used to determine temperatures reached by the teeth and mouth of piglets during resection by grinding. As the area under investigation is small it is also important to test whether the IRT applied are able to provide data usable in this application. The procedure was performed by an experienced operator using an electric grind stone on twelve piglets 16 hours after birth. IRT video was taken at 30 frames per second with thermal imaging camera model TVS500; different distances and camera angles from the mouth of piglets were checked. It was demonstrated that if resection of teeth by grinding is correctly performed the temperature (T) rise is confined to the teeth themselves and does not extend to any other part of the mouth. The average baseline T of the mouths is 37 °C. The teeth grinding procedure took on average 50 seconds per piglet to accomplish. The time grinders were applied to each tooth was less than 2 seconds. Immediately on application of the grinder the average tooth T reached was 50 °C remaining virtually constant during the 2 seconds. The peak temperature of a tooth reached was 88 °C for a sub-second interval. Teeth took 2 seconds on average to cool back to baseline temperature. It is actually possible to record a IRT video during the grinding of the teeth of piglets without interfering with the procedure and collect valuable useable data. However, it is imperative to use a camera with spatial resolution at least equal to that of the TVS500, to ensure the accuracy of the T values obtained on small objects such as teeth. The distance of 0.4 m between camera and subject has proved to be the optimal one and the frequency of 30 frame per second appropriate. It would seem that, with proper handling, grinding of teeth does not cause heating to the softer tissues of the mouth and only effects the tooth for a very short period of time. Further behavioural and physiological examinations would need to be carried out in a larger study to assess the overall effect of grinding on the welfare of the piglet and to further investigate the debate on the most suitable procedure for tooth resection of piglets. This is a very new and unique measure which supplies great detail and could be pivotal in understanding the true effects of such procedures on piglets.

Demonstration of thermoregulatory control of piglets during farrowing by infrared thermography

Redaelli, V.[1], Farish, M.[2], Luzi, F.[1] and Baxter, E.M.[2], [1]Univ. of Milano, DSA, via G. Celoria 10, 20133 Milano, Italy, [2]SAC, Behaviour and Welfare, Sustainable Livestock Systems, West Mains Road, EH9 3JG Edinburgh, United Kingdom; fabio.luzi@unimi.it

Newborn piglets are extremely cold sensitive, being born with very little adipose tissue, no brown fat and very little, if any pelage. They lose heat rapidly once born and need to get to the udder and ingest colostrum quickly in order to increase their body temperature, as well as acquire immunity. The purpose of this study was to monitor piglet' skin temperature continuously during farrowing, to better demonstrate the thermal compromise piglets face once born, without the need to disrupt maternal and piglet behaviour during farrowing by removing piglets for traditional methods of measuring temperature (e.g. rectal thermometers). A preliminary study was conducted within an existing research project on alternative farrowing systems. Sows gave birth in loose farrowing accommodation onto a solid concrete floor with a designated amount of straw bedding. The infrared thermal imaging camera, model Avio TVS 500 (thermal resolution better than 0.05 °C, spatial resolution 1.6 mrad), was secured two metres above the nest pen at the time when a sow was beginning to show nest-building behaviour. A thermal image of the nest, sow and subsequent piglets was captured automatically every 20 seconds starting seven hours before and until 12 hours after the farrowing. Images obtained were analyzed using software specialized in processing infrared images (Goratec Thermography Studio), A thermographic video showing piglet' skin temperature variation was created. Piglet' skin temperature decreased from a maximum value of 39.5 °C immediately at birth to just 31-32 °C ten minutes after birth, which was equal to the temperature of the straw near the sow. Concrete flooring not covered by straw bedding close to the nest varied in temperature between 18-20 °C. Piglet' skin temperature started to increase only when they reached the udder and started suckling; taking more than one hour after birth to reach 35 °C. To our knowledge, this is the first study using infrared thermography to continuously measure changes in skin temperature on pigs during farrowing. Thermography could provide highly detailed and non invasive measures on the multifactorial influences on piglet thermoregulation in the most critical time post birth and lead to an improvement in knowledge of the behavioural and physiological events that occur during this period. Thermography may better demonstrate the importance of micro-climate to buffer the piglet from immediate heat loss and therefore potentially give it extra time to reach the udder and suckle colostrum. In addition, this technique offers a useful, visual tool to translate scientific data into a visual medium for the producer.

Correlation between stepping and heart rate variability in dairy cow at milking

Speroni, M. and Federici, C., Agricultural Research Council (CRA), Fodder and Dairy Productions Research Centre (FLC), via Porcellasco 7, 26100 Cremona, Italy; marisanna.speroni@entecra.it

Restlessness (kicks and steps) as well as heart rate variability (HRV) have previously been used together or separately to assess stress in dairy cows at milking. There is some debate in the field of animal welfare assessment regarding the interpretation and use of parameters derived from the frequency domain analysis of HRV; however, it is generally accepted that high frequency power band (HF) indicates vagal activity, low frequency power band (LH) is associated with both sympathetic and vagal acitivity, LH /HF ratio is a measure of the sympatho-vagal balance and the LF/HF ratio increases with sympathetic dominance, indicating a stress response. Aim of this study was to better understand relationship between behaviour and HRV as animal welfare indicators at milking. Twenty-four dairy cows were monitored at milking; behaviour recordings were done by direct observation at evening milking for two consecutive days; movements of legs were recorded continuously from the attachment of teat cup to the end of milking; stepping was defined as a cow shuffling its hind feet. Heart rate variability was measured using a commercial heart rate monitor (Polar®); HRV data concerning the first 5 minutes of milking was analysed by Kubios software (Dep. of Physics, Kuopio Univ., Finland). Since respiration rates affect the location of HF band, an average respiration rate of 12-35/min was considered to set HF limits at 0.20 and 0.58 as suggested by other authors for cattle. Stepping resulted moderately correlated with LH/HF ratio ($R=0.5$ $P<0.05$); an higher number of steps corresponded to an increased LF/HF ratio indicating an acute change in the sympato-vagal balance; an homeostatic adjustment and a vagal modulation in response to sudden changes in body position or movements were expected, however, since only 25% of variability in LF/HF ratio seemed to be explained by the effect of movement, we concluded that LF/HF ratio gave additional information to that provided by the number of steps alone thus it is worth to associate HRV to behavioural measures when stress at milking needs to be assessed.

Animal based parameters from the farmers' point of view – results of a pilot study on the implementation of herd health and welfare plans in german organic dairy herds

Brinkmann, J.[1], *March, S.*[1] *and Winckler, C.*[2], [1]*Georg-August-University of Goettingen, Department of Animal Sciences, Location Vechta, Driverstrasse 22, 49377 Vechta, Germany,* [2]*University of Natural Resources and Life Sciences, Department of Sustainable Agricultural Systems, Gregor-Mendel-Strasse 33, 1180 Vienna, Austria; jan.brinkmann@agr.uni-goettingen.de*

Herd health and welfare plans are a tool to monitor and improve animal health and welfare in livestock farms. Experiences with British herd health plans show that the acceptance of such plans by the farmers is an essential prerequisite for its successful transfer into practice. Animal-based parameters of health and welfare form an integral part, but only little is known how well these are perceived by the farmers in the planning process. Within a pilot study on the implementation of herd health and welfare plans in German organic dairy farms it was therefore the aim to investigate farmers' expectations toward herd health and welfare plans and the animal-based indicators farmers are familiar with including target and intervention levels. 40 organic dairy farmers (average herd size 70 cows) were interviewed at the start of the study. Three years after the plans had been implemented in 27 farms, these farmers were interviewed again, investigating whether expectations had been met and evaluating the acceptance of the animal based parameters applied in the planning process. Before the start of the pilot study, the most frequently mentioned animal-based measures were milk somatic cell count (n=37/40) and lameness prevalence (n=31/40). Target values provided by the famers (SCC: median 200.000; 100.000-400.000; lameness: median 7.5%; 0-25%) were within the range of those obtained from experts. Three years later, data from records such as from milk recording (n=27/27) and treatment records (n=10/27) were regarded important, but also parameters which have to be assessed directly in the animals such as locomotion (n=13/27), body condition (n=12/27) and skin lesions on the limbs (n=3/27). The organic dairy farmers also had a positive attitude toward herd health and welfare plans. They highly valued the strategic approach of herd health and welfare planning with focus on animal-based parameters (16/27 grade 1 (very important), 8/27 grade 2 (important) and 3/27 grade 3 (more or less important)). In conclusion, the present results indicate that the use of animal based parameters in the course of health and welfare planning in organic dairy farming is feasible and well accepted by organic milk producers. Animal based parameters were highly important for the farmers and this finding calls for the further inclusion of animal based parameters in preventatively orientated concepts to improve animal health and welfare.

Improving welfare for dairy cows and calves at separation

Johnsen, J.F.[1], Grøndahl, A.M.[1], Ellingsen, K.[1], Bøe, K.E.[2], Gulliksen, S.M.[3] and Mejdell, C.M.[1], [1]Norwegian Veterinary Institute, Section for disease prevention and animal welfare, P.O. Box 750 Sentrum, N-0106 Oslo, Norway, [2]Norwegian University for Life Sciences, Department of Animal and Aquacultural Sciences, P.O. Box 5003, N-1432 Ås, Norway, [3]Norwegian School of Veterinary Science, Department of Production Animal Clinical Sciences, P. O. Box 8146 Dep, N-0033 Oslo, Norway; ann-margaret.grondahl@vetinst.no

Introduction In conventional dairy production cow and calf are usually separated within hours after birth, whereas according to rules for organic dairy production in Norway and Sweden it is mandatory to keep cow and calf together for at least three days, and the calves have to be fed whole milk for 90 days. Separation of cow and calf after a period of free suckling breaks the established bond causing behavioural reactions. The intensity and extent of these reactions reflect distress and hence reduced animal welfare. Allowing physical contact after separation has been shown to alleviate behavioural responses of beef cattle. The present study was performed to compare the behavioural reactions of cow and calf of dairy breed (Norwegian Red) following separation with or without physical contact. Materials and methods The study compared two separation methods after 7-8 weeks of free suckling: fence-line separation (FL, n=12 cow-calf pairs) where animals could have physical contact or separation merely with auditory contact (A, n=12 cow-calf pairs). At the day of separation, the calves were locked in their calf creep located adjacent to the cows housed in deep straw loose housing. The barrier between the cows and calves after separation consisted either of an open fence (FL) or an opaque, two meter high wall (A). Two cow-calf pairs were separated at a time and the test animals were allocated to one of the two treatments. Behaviour of cow and calf was recorded by manual observation using a combination of instantaneous recording every 5th minute and continuous recording for 2×2 hours at day 0, 1, 2, 3 and 4 after separation. Calves were bottle fed 3×2 litres of milk per day after separation and were fed just before observation periods. Results Preliminary analyses show that FL significantly reduced high pitched vocalization in calves ($P<0.001$), resulted in less alert behaviour including restless walking and more time lying ($P<0.001$). Among cows there were no differences in vocalisation but FL cows rested and ruminated more ($P<0.001$). Conclusions Separation methods allowing physical contact between the cow and calf can increase animal welfare in dairy herds practicing suckling systems.

Innovative animal-based measures for monitoring the welfare of broiler chickens

Tuyttens, F.A.M., Vanderhasselt, R. and Buijs, S., Institute for Agricultural and Fisheries Research (ILVO), Animal Sciences Unit, Scheldeweg 68, 9090 Melle, Belgium; frank.tuyttens@ilvo.vlaanderen.be

Indicators for monitoring farm animal welfare should be valid, reliable, and feasible. Moreover, there is increasing consensus that preference should be given to animal-based indicators because of their more direct relationship with animal welfare than housing or management parameters. For broiler chickens there seems to be a shortage of indicators that meet all these criteria. As a consequence, current protocols for monitoring the welfare of broilers may have limited sensitivity and/or fail to cover certain aspects of animal welfare. It seems that welfare problems that do not cause obvious physical lesions are particularly hard to detect. We provide an overview of recent research conducted at the ILVO on the development and validation of several innovative animal-based measures for monitoring broiler welfare. One-dimensional measurements of fluctuating asymmetry (FA) have been shown to increase after exposing chickens to some stressors. Furthermore, elevated levels of FA induced by cold-stress during early life persisted until slaughter-age, suggesting that FA at slaughter-age may reflect the cumulative effect of stressors during the broiler's entire lifespan. Shape asymmetry based on 2- and 3-D measurements may be even more sensitive to stress, but the feasibility of applying FA measurements in the context of broiler welfare monitoring programs remains problematic. Absence of prolonged thirst is considered one of the most important aspects of animal welfare, but is usually assessed by resource-based measures such as the number of animals per drinker. The sensitivity and validity of these measures are questionable. As alternative tests of thirst, we present an on-farm test based on spontaneous water consumption rate from an unfamiliar and easy-to access drinker, and an at-slaughter test based on total blood volume. Society is very concerned about the restricted space allocated to farm animals. Current animal welfare monitoring protocols usually assess spatial requirements by estimating the stocking density. We have investigated novel indicators of crowding based on the observed versus expected distribution of broilers over the available space, their willingness to work (i.e. cross a barrier) for access to a less crowded area, and the rate at which a vacated area is replenished. The latter test shows most promise for on-farm application. Several animal-based measures presented here may be included in broiler chicken welfare assessment protocols in order to improve their sensitivity and completeness, although further refinement and validation are warranted.

Analysis of dynamics of aggression as a tool to improve group housing for rabbit does

Rommers, J.M., Gunnink, H., Klop, A. and De Jong, I.C., Wageningen UR, Livestock Research, Animal Welfare, P.O. Box 65, 8200 AB Lelystad, Netherlands

Group housing of rabbit does in commercial production full fills the animals basic need for social contact. Until now aggression among does is considered the main bottleneck to put group housing of does into practice as it causes wounds and increased mortality. Previous research indicated that, besides installing the hierarchy in a group, also the lay-out of the group housing system currently used possibly caused aggression. Therefore, in the present study aggression between does was studied in detail in order to improve the group housing system. Thirty-two hybrid rabbit does (Hycole) were housed in groups of eight from 15 d after parturition until the next parturition in wire cages with an elevated platform that were connected to each other using holes in the side- walls. Does were housed individually in the same system during the first 14 d after parturition by closing the holes. Video-recordings were made for 24 h on the first and the third day that the does were housed in groups. The number of aggressive and social interactions were scored. The aggressive interactions consisted of offensive (threatening, attacking, fighting, chasing) and defending (fleeing, submitting, withdrawing) behaviors. Also the does involved and the place in the system were aggression occurred (own or foreign cage, elevated platform or bottom of the cage) were noted. It turned out that holes were frequently blocked by other does which hampered fleeing behavior. The total number of agonistic interactions in a group was high, although a decrease was observed from d 1 (148± 24) to d 3 (51±27). On both days 45% of the observed interactions was offensive, 30% defensive and 25% social. On both days, 84% of the offensive behavior consisted of attacks and fights. On d1 69% of the defensive behavior consisted of flight, where on d3 flights and withdrawals both counted for almost 50%. Submissive behavior was nearly observed (0.5%). Offensive behavior was observed throughout the system, whereas the defensive behavior was mainly expressed in foreign cages at the elevated platforms. All does in a group were involved in agonistic interactions on d1, although frequency differed between individuals. On d3 only a few does were causing aggression in a group. Does predominantly showed offensive behavior in or in the neighborhood of what was previously their own cage. It can be concluded that the cage lay-out seemed inappropriate due to lack of space, flight possibilities, and places to hide from conspecifics. Moreover, does still discriminate between their own and a foreign cage after being housed in a group.

Assessment of mechanical pain sensitivity in the pig's tail

Di Giminiani, P.[1], Herskin, M.S.[1] and Viitasaari, E.[2], [1]University of Aarhus, Department of Animal Health and Bioscience, Blichers Allé 20, 8830 Tjele, Denmark, [2]University of Helsinki, Faculty of Veterinary Medicine, P.O. Box 66, 00014 Helsinki, Finland; pierpaolo.digiminiani@agrsci.dk

In many countries, piglet tails are docked within the first week of life in order to prevent tail biting. This procedure leads to the formation of traumatic neuromas in the tails, suggesting that the pigs experience increased pain sensitivity and possibly spontaneous pain. At present, however, no assay is available for on-farm quantification of porcine pain sensitivity in tails. As part of the establishment of proper methodology for assessment of mechanical pain sensitivity in the tails of growing pigs, we performed a 2 × 2 factorial experiment involving the following comparisons: 1) gender of the pigs (females vs. castrated males); and 2) familiarity with the testing environment (tested on first visit vs. tested after 3 visits to the testing environment over a 3 day period). 12 males and 12 females growing pigs of 57-70 kg, and kept in commercial slaughter pig conditions were used as experimental animals. The nociceptive mechanical stimulation was applied using an electronic Von Frey anesthesiometer and directed at the mid-section of the tail, which was stabilised by use of a PVC transparent tube with a top opening. The pigs were always tested with a companion animal present, and the testing took place in a 3 × 3 m test arena surrounded by wooden walls and situated in a room in the same building as the home pens of the animals. Each pig received 4 consecutive stimulations and a mean mechanical threshold was calculated. For all stimulations, the experimental animals responded with a tail flick within the maximum time limit (30 s). Neither familiarity with the test arena (365±155 and 380±163 g respectively, independent t-test: $P>0.05$) nor gender (370±158 for males and 375±182 g for females, independent t-test: $P>0.05$) did have significant effects on the mechanical pain sensitivity on the tails. The present experiment is among the first to focus on quantification of pain sensitivity in tails of growing pigs, and the results might suggest that this method can be applicable for on-farm quantification – a situation where it would be needed to remove the animals from the home pen, but where application of standardized habituation protocols might be problematic. However, further validation studies are needed to verify this.

Is the response to humans consistent over productive life in dairy cows?

Haskell, M.J.[1], Bell, D.J.[1] and Gibbons, J.M.[2], [1]SAC, Sustainable Livestock Systems, West Mains Road, EH9 3JG, Edinburgh, United Kingdom, [2]Agriculture and Agri-Food Canada, P.O. Box 1000 6947 #7 Highway, Agassiz, V0M 1A0, British Columbia, Canada; marie.haskell@sac.ac.uk

Dairy cattle have a high level of interaction with humans. If cows find these interactions stressful or aversive, this can affect their wellbeing and productivity. Because of this, the quality of response that cows show towards humans is a parameter that should be included in welfare assessment tools. A common test of responsiveness or fear used in welfare assessment protocols for dairy cattle is the human approach test (HAT). It involves approaching a subject animal slowly and scoring the point at which she begins to move away. For the HAT to be used to assess welfare on farms, it must be robust and reliable. This includes determining whether the response to the HAT depends on the age of the cow tested, as different cow age profiles on farms may bias the farm-level score. If the response stabilises at a certain age, an appropriate sampling strategy can be devised. To determine whether the HAT response varies with the age, 122 Holstein cows were tested at regular test intervals across their productive lifetime. The first test stage for heifers was at 12-15 months of age, then at first breeding and then prior to their first calving. The animals were then tested in early (30-50 days in milk (DIM)), mid- (130-150 DIM) and late (230-250 DIM) lactation for their first and second lactations and finally early in the 3rd lactation. The test involved approaching the cow when she was standing in the passageway of her home pen with sufficient space around to move away. The response was recorded on a scale of 0 to 8 where a low score indicated a responsive or fearful animal. Descriptive qualitative terms were also used to capture the quality of response. Boldness, at ease and fearfulness were scored using sliding scales from absence to full presence. REML was used to analyse all the scores. There was a significant effect of test stage on HAT response score ($P<0.001$). Cows became more approachable with increasing age, up until the middle of the first lactation. There was no further change in the scores beyond this stage ($P=0.205$). There was a significant effect of test stage on the qualitative terms, with cows becoming bolder, more at ease and less fearful with increasing age ($P<0.05$ for all). For the individual cow, her ranking within the group at each stage showed good correlation with the rankings in the following stages, suggesting within-individual consistency. The results suggest that the HAT is a robust test, and that it can be used successfully to compare responsiveness or fearfulness across farms if the scores are taken from cows which are in the middle of their first lactation or older.

Comparison of different behaviors as indicators of distress in piglets euthanized via CO_2 or mixed CO_2:Argon gas at different flow rates using the Smart Box euthanasia device

Sadler, L.[1], Hagen, C.[2], Wang, C.[1], Widowski, T.[3] and Millman, S.[1], [1]Iowa State University, 1600 S 16th St, 50011 Ames, IA, USA, [2]Value-Added-Science & Technologies, 2393 McDonal Ave, 50401 Mason City, IA, USA, [3]University of Guelph, 50 Stone Road E, N1G 2W1 Guelph, Ontario, Canada; smillman@iastate.edu

In this study we examined different behaviors as potential indicators of distress in piglets during gas euthanasia. A CO_2:Argon (CA) gas mixture or 100% CO_2 gas were applied to weaned piglets 16 to 24 days of age. A total of 180 piglets, BW 4.6±0.7 kg, were utilized. Two gas mixtures (100% CO_2 and 50:50 CA) and 4 flow rates: slow (SL), medium (MD), fast (FT), and prefill (PF); 20%, 35%, 50%, and prefill with 20%, chamber volume per minute respectively, were examined. A control treatment (CT) passed ambient air through the chamber followed by blunt force trauma. A barrow and gilt were placed in a plastic chamber with the lid and one side composed of clear plastic to facilitate behavior observations. A Smartbox device (Euthanex Corp, Palmer, PA) was used to supply gas at controlled rates. Piglets were scored using direct observation for latency to perform 3 behaviors associated with insensibility: last movement (LM), loss of posture (LP), and gasping (GSP) Six additional behaviors: open mouth breathing (OMB), licking and chewing (LC), nasal discharge (ND), defecation (DEF), urination (UR) and vomiting (VM), were scored as zero/one data to give percentage of piglets displaying behaviors. LM data was log transformed and analyzed using a mixed model with fixed effects of treatment, and blocked by day of treatment. All other behaviors were analyzed using univariate product-limit estimation of the survival curves, with a post hoc adjustment for multiple comparisons. Significance was determined at $P<0.05$. Differences were observed for behavioral indicators of insensibility. Latency (seconds) to LM was shortest for PF & FT, followed by MD then SL (269, 274, 313, and 529 respectively). LP and GSP followed a similar pattern (means ranged over all 4 flow rates LP: 97- 200, GSP: 46-159). All other behaviors were observed with the exception of VM. A difference ($P<0.05$) was observed for OMB. No piglets displayed this behavior in the CT, while the gas treatments ranged from 80 – 100%; differences were not observed between the gas type or flow rates for OMB. No differences ($P>0.1$) were observed for any other measures: LC (5-70%), ND (0-30%), DEF (25-60%), UR (5-35%). In conclusion, CA and SL prolonged the duration to insensibility, as measured by LM, LP, and GSP. When examined as percentage of animals displaying behaviors, OMB was the only behavioral measure to discern differences between treatments, but only for ambient air versus all other gas treatments.

A preliminary study on the assessment of pain induced by mutilation in piglets

Lonardi, C., Brscic, M., Scollo, A. and Gottardo, F., University of Padova, Department of Animal Science, viale dell Università 16, 35020 Legnaro Padova, Italy; chiaralo99@yahoo.it

Aim of this preliminary study was to assess differences in posture and walking in 4-6 days old piglets before and after procedures of castration and/or tail docking in the wider objective to study efficacy of pain relief treatments. A total of 59 piglets (25 females and 34 males) belonging to 5 litters of 10-13 piglets each were used. All piglets were handled in the same way; females were only tail docked while males were also castrated according to the farm common practices. Expecting that an animal in pain assumes a different stance and changes walking behaviour in order to relief pain, differences between pre- and post- castration and/or tail docking were measured. Each piglet was visually inspected by a trained veterinarian when standing and walking in the farrowing crate 1 hour before (T -1), right after (T 0), and 1 hour after (T +1) the surgical interventions. Occurrence of weaker and protracted forward hind limbs, hind limb nonweightbearing, hind tiptoe walking, and kyphosis were recorded during direct observation on a three point scale from 0 to 2 (0 = correct posture, 1 = slight alteration, 2 = severe alteration of the posture). Moreover, piglets' paws prints were taken at T -1 and at T +1 in order to evaluate change in the steps length. Data regarding vet evaluation were processed adopting the proc logistic of SAS whereas data on distances among paws prints were submitted to ANOVA using a mixed model which considered as fixed effects sex, observation (before and after surgical intervention) and their interactions whereas the piglet within litter and sex was the random effect. No differences between males and females were detected before tail docking and castration for occurrence of weaker and protracted forward hind limbs, hind limb nonweightbearing, hind tiptoe walking, and kyphosis. After surgical procedures, only castrated male showed changes in the locomotory behaviour even if score 2 has been used only in the case of animals that appeared weaker before surgical intervention. Score 1 for hind tiptoe walking after castration had a higher risk of occurrence after castration (OR: 9.3 $P<0.05$ at T 0, and OR: 12.9 T +1 $P<0.001$). Kyphosis were also affected by castration in the observation carried out 1 hour after (OR: 24.0, $P<0.001$). The steps' lengths were not significantly modified in the females group after tail docking and in the males group after castration. The observation of locomotory behaviour by a trained observer seems to be a useful indicator of pain in piglets after castration and, considering that the method is easy to apply and does not require an invasive manipulation of the animals, it can be therefore repeated in order to monitor pain over time.

Welfare and behaviour in an aviary laying-hen housing system

Parsons, R.[1], Hayes, M.[2], Xin, H.[2] and Millman, S.[1], [1]Iowa State University, Veterinary Diagnostic & Production Animal Medicine, 2412 LVMC, 50011, USA, [2]Iowa State University, Agricultural & Biosystems Engineering, 3204 NSRIC, 50011, USA; bparsons@iastate.edu

Our objective was to assess the use of a litter area in a multi-tier aviary laying hen facility in relation to changes in welfare assessment parameters. Two Iowa aviary hen houses were used in this study, consisting of rows of 3-tiered enriched cages and litter areas that birds had access to from 12:00 to 21:00 daily. The rows were divided into 15.2 m sections down the length of the barn, housing 1,500-1,700 Hy-Line Brown hens per section for a total of 50,000 hens per house. Ten sentinel sections were selected for welfare assessment and behavior observations. A modified Welfare Quality assessment was performed at peak, mid and end of production for each of the 10 sections, selecting 10 hens from each of the sections for individual hen scoring. Clinical scoring of bird health (plumage, parasites, injuries, disease) and behaviour (novel object test [NOT], avoidance distance test [ADT]) were performed. Four video cameras were mounted in a section to capture video images of the lowest cage tier, from which hens accessed the litter area. During peak through mid-production laying periods, cameras were rotated biweekly amongst the 10 sentinel sections, such that each section was filmed continuously during the light hours. To determine litter use, frequencies of hens leaping to and from the litter area were determined using 10-minute continuous sampling from 12:30 to 20:30. Data were analysed using mixed linear models, and preliminary results for peak and mid-production assessment visits are presented. Avoidance distance did not differ between visits (29.7±10.4 cm, P=0.71). At peak-production, all hens had excellent plumage (score=0), but at mid-production 27% of hens had a score of 1 and 3% had a score of 2. Keel deformities were found on 6% and 12% of birds at peak and mid-production, respectively. Approximately, 154% of the birds went to the litter during the day, indicating that some hens visited the litter and returned to the tiers more than once a day. Movement of hens to (P=0.02) and from (P=0.014) the litter area was affected by time of day, with more hens moving during the mid-afternoon (14:30 to 15:30) and evening (19:30) time periods. There was also a significant difference between sections of the house for frequency of movement to the litter (P=0.02), but it is unclear if this represents differences due to group or due to week. In conclusion, some welfare assessment parameters, such as keel injuries, changed over time within the same group of hens, and further research is needed to determine risk factors for correction. Litter was a valuable resource for these hens that, on average, accessed the litter area more than once daily.

The care and welfare of horses destined for slaughter: recommended handling guidelines and animal welfare assessment tool for horses

Woods, J.[1]*, Stull, C.*[2] *and Grandin, T.*[3]*,* [1]*Horse Welfare Alliance of Canada, Box 785, Cochrane, AB T4C 1A9, Canada,* [2]*University of California, Davis, California, USA,* [3]*Grandin Livestock Services, Fort Collins, Colorado, USA; livestockhandling@mac.com*

Horse processing is an economically feasible end of life option for horse owners. Industry working through the Horse Welfare Alliance of Canada believes that in order to ensure the well being of our animals we must provide viable end of life options. Horse processing is a legal and viable industry in Canada – in fact, horsemeat is our third largest exported meat. Horse owners, caregivers, handlers, enforcement personnel, regulators, animal activist and the equine industry all have a role to play in ensuring horses are treated humanely and with respect throughout their lives. This includes on farm, during transport and at end of life, including processing. The Recommended Handling Guidelines and Animal Welfare Assessment Tool for horses destined for processing provides guidelines and training for all those involved in the processing industry to ensure the animals are receiving the best care possible. The guideline and assessment tool was developed in consultation with leading animal welfare scientists, equine behaviorists and horse slaughter experts from across North America. The guidelines offer detailed information about equine behavior and handling, facility design for optimal animal welfare at loading, unloading, lairage and within plant handling areas, transportation, and proper stunning, along with animal welfare assessment standards and forms. The benefits of this guideline include consistent industry guidelines for the care and handling of horses destined for processing, including: an animal welfare management tool for both the industry and third party assessors; a standardized tool to assess the welfare of horses during all stages leading up to and including processing; concise standards for what is acceptable, positive messaging to the public and customers that this industry is responsible and care about the animals, and document factual information on the processing of horses in North America.

Can feather scoring be used to assess the degree of chronic hunger in broiler breeder hens?

Morrissey, K.L.H.[1], Widowski, T.M.[1], Leeson, S.[1], Sandilands, V.[2], Classen, H.[3] and Torrey, S.[1,4], [1]University of Guelph, Animal & Poultry Science, 50 Stone Rd East, N1G 2W1 Guelph, Canada, [2]Scottish Agricultural College, Avian Science Research Centre, Auchincruive, Ayr, United Kingdom, [3]University of Saskatchewan, Animal & Poultry Science, 51 Campus Dr, S7N 5A8 Saskatoon, Canada, [4]Agriculture & Agri-Food Canada, 93 Stone Rd West, N1G 5C9 Guelph, Canada; kmorriss@uoguelph.ca

Due to their capacity to grow quickly, broiler breeders must be severely feed restricted to maintain healthy body weights. This restriction reduces welfare and can induce stereotypic behaviour, including feather pecking, which has negative implications for both the pecker and victim. It has been suggested that the problem may be symptomatic of chronic hunger or a lack of dietary fibre or foraging substrate. By measuring levels of feather pecking, one aspect of welfare can be assessed. This study determined whether feather pecking could be reduced and welfare improved via dietary manipulation. There were 6 treatment diets, each with 5 replicate pens of 9-12 birds. Control diets consisted of a commercial crumble, fed on a daily or skip-a-day (SAD) basis. Treatment diets included soybean hulls as a bulking ingredient and calcium propionate as an appetite suppressant of either a feed grade (FG) or purified (P) quality. Both treatment diets were also fed on either a daily or SAD basis. Approximately half (5 or 6) of the birds were randomly chosen from every pen and were feather scored at 10, 14, 20, 26 and 36 weeks of age. Six body parts (neck, back, wings, legs, vent area, tail) were given a score from 0-5 (0 = no feather damage, and 5≥50% feather loss with tissue damage). Scores were summed for each bird, and averaged for each pen. Data were analyzed with Room and Feeding Frequency (daily, SAD) as main factors and Diet (control, FG, P) as the sub-factor with repeated measures (SAS 9.2). There was an effect of Diet ($P<0.01$), with the control birds having worse feather scores compared to the combined means of the two alternative diets. Feeding Frequency did not affect feather score ($P=0.25$). However, the interaction between Frequency and Time was significant ($P=0.01$), with SAD-fed birds scoring better than daily-fed birds at 20, 26 and 36 weeks of age. This interaction could indicate that the SAD regime increased satiety after the birds became accustomed to the schedule. These differences indicate that an increase in satiety (one measure of welfare), may be determined by assessing feather coverage. Since feather pecking is often seen in combination with other stereotypies, a reduction in one may be indicative of an increase in overall welfare. Therefore, feather scoring may be a useful tool for producers to assess the level of hunger in broiler breeder flocks.

Prevalence of sickness behavior in neonatal dairy calves at a commercial heifer facility and associations with ADG and illness

Stanton, A.L.[1], Widowski, T.M.[1], Kelton, D.F.[1], Leslie, K.E.[1], Leblanc, S.J.[1] and Millman, S.T.[2], [1]University of Guelph, 50 Stone Road, Guelph, Ont, N1G 2W1, Canada, [2]Iowa State University, 1600 S 16th Street, Ames, Iowa, 50011, USA; astanton@uoguelph.ca

Recognition of key behaviours associated with common diseases and poor growth can allow new and experienced animal caretakers to improve identification of potentially sick animals. The objectives of this research project were to determine which behavioural measures are associated with reduced growth, and if these behaviours can be used to identify ill animals. Holstein heifer calves (n=744) were housed in individual pens in a naturally-ventilated nursery barn at a commercial heifer rearing facility in western New York. Weaning was initiated at 5 wks and completed at 6 wks of age. Calves were monitored for growth and disease between 0 and 8 weeks of age by barn staff. Behavioural observations were performed every other week with each calf observed during three time periods. Calves were 10±5 (mean ± SD), 24±5, and 38±5 days of age at observation periods 1, 2, and 3, respectively. The observer scored calves based on 5 behavioural tests; lying position, standing posture, vigilance, human approach test, and lethargy test. The probabilities of each behaviour occurring over time were analyzed using generalized linear mixed models with a logit transformation. Models of ADG were evaluated using a generalized linear mixed model with random effects for source farm and enrolment group. At observation 1, the behavioural responses of a subset of calves with active diarrhea (n=229) were compared to calves with no active disease (n=47). Differences in behavioural responses were analyzed using chi-square analyses. The ADG of calves that were observed statue standing at observation 1 was 0.11±0.03 kg lower than calves that were observed with a normal posture (P=0.002). Diarrheic calves were more likely to be short lying than their healthy cohorts (P=0.04). A high lethargy score at observation 3 was associated with lower ADG (P=0.006). The proportion of calves that approached the observer during the human approach test was 46% (105/229) and 70% (33/47) in the diarrhea and control groups, respectively (P=0.0007). In conclusion, short lying, although uncommon, appears to be a highly specific indicator of diarrhea. Also, behavioural tools such as lethargy score, statue standing and willingness to approach the observer may be useful to identify animals with decreased average daily gain and diarrhea. These behaviours could be integrated into future studies of predictors of disease and discomfort, as well as used on farm for training of personnel in disease detection.

Use of pedometry for detection of lameness caused by digital dermatitis in dairy cows

Higginson, J.H.[1], Millman, S.T.[2], Cramer, G.[1,3], Leslie, K.E.[1], De Passille, A.M.B.[4], Duffield, T.F.[1] and Kelton, D.F.[1], [1]University of Guelph, Population Medicine, 50 Stone Rd E, Guelph, ON N1G 2W1, Canada, [2]Iowa State University, Veterinary Diagnostic & Production Animal Medicine/ Biomedical Sciences, College of Veterinary Medicine, 1600 S. 16th St., Ames, Iowa 50011, USA, [3]Cramer Mobile Bovine Veterinary Services, 144 Oak Street, Stratford, ON N5A 7L5, Canada, [4]Agriculture and Agri-Food Canada, 6947 #7 Highway, P.O. Box 1000, Agassiz, BC V0M 1A0, Canada; jhiggins@uoguelph.ca

Lameness is one of the major welfare concerns in dairy cattle, and is difficult to resolve due in part to the poor detection rates. The objective of this pilot study was to examine changes in dairy cow activity around lameness events, a component of a larger ongoing investigation to determine the efficacy of pedometric activity for early lameness detection. The commercially-available Pedometer Plus activity monitoring system for cattle (SAE Afikim, Israel) provides the number of steps taken, the duration of lying time, and number of lying bouts. Pedometers were affixed to the left hind limbs of 130 lactating cows. Hooves were examined for lesion identification approximately every 3 months and trimmed every 6 months for a 14 month period from June 2009-Augst 2010. In addition, lameness cases throughout this time period were identified by the producer were evaluated by a veterinarian and treated with oxytetracycline. Preliminary results of five cows with new cases of digital dermatitis have been examined. Activity and lying behaviour were analyzed during two time periods – seven days prior to lameness exam, and seven days following exam, with exam day excluded. A paired t-test demonstrated a difference in activity between time periods (P=0.03). On average (± sd), cows took 65.1 (± 4.1) steps/ hour prior to lesion identification and 76.2 (± 5.6) following identification. The mean number of lying bouts between time periods was not different (P=0.79), with mean lying of 10.0 (± 0.7) bouts/day prior to the lameness exam and 9.9 (±0.6) following the lameness exam. Lying duration also did not differ (P=0.54), with 693.1 (± 28.0) minutes/day prior to the lameness exam and 675.1 (± 34.5) minutes/day following the exam. These preliminary results suggest a significant difference in activity between the week prior and following lameness identification and treatment. Data will also be analyzed to determine if pedometry is more or less effective than farmer subjective scoring. Additionally, continued data analysis of greater periods of time surrounding lameness events and of other lesions will determine if early identification with pedometry is possible.

A training program to ensure high repeatability of injury and body condition scores of dairy cows

Gibbons, J., Vasseur, E., Rushen, J. and De Passille, A.M., Agriculture & Agri-Food Canada, Agassiz, V0M 1A0, Canada; jenny.gibbons@hotmail.co.uk

Obtaining reliable animal-based measures from commercial farms can be challenging. We developed a training program to train naïve observers on animal-based measures as part of an on-farm welfare assessment of dairy cows. Inter-observer repeatability with the trainer was evaluated during training on three animal based measures (hock and knee injury using a 4-point scale and body condition score using a 5-point scale (BCS)). Four observers completed a 1 week-training session which started with 2 h classroom and 2 h live instruction sessions (d1). The classroom session included tutorials on Standard Operation Procedures (SOP), a discussion of 8 sample pictures and independent scoring of hock (n=20), knee (n=20) and BCS (n=12) pictures. A BCS CD-Rom and a flow chart were provided to assist in BCS training. During the live session, observers scored injuries and BCS on cows (n=20). The following day, the observers scored the same cows a second time (+1 d). Six and seven days after d1, observers scored 20 cows each on two commercial farms (+6 d, +7 d). Percent exact agreement and weighted kappa were calculated between observers and the trainer. On d1, agreement ranged from low to high for hocks (pictures= 55-70%, kappa 0.55-0.67; live= 65-79%, kappa 0.56-0.82) and knees (pictures= 40-73%, kappa 0.02-0.57; live= 49-63%, kappa 0.22-0.46). Between +1 d and +6 d agreement reduced but improved again on +7 d for hocks (+1 d= 70-94%, kappa 0.64-0.90; +6 d = 59-77%, kappa 0.32-0.56; +7 d= 84-89%, kappa 0.61-0.80) and knees (+1 d= 77-89%, kappa 0.53-0.67; +6 d= 53-67%, kappa 0.19-0.33; +7 d= 68-87%, kappa 0.44-0.81). Agreement was high for BCS on the live training and remained moderate to high over time (pictures= 42-58%, kappa 0.04-0.31; live=92-100%, kappa 1.00-1.00; +1 d=82-100%, kappa 0.86-1.00; +6 d=79-100%, kappa 0.40-0.66; +7 d=95-100%, kappa 0.86-1.00). The intensive training program achieved repeatability of 84% and kappa >0.6 for hock injuries and BCS over a 1 week period. The BCS SOP had higher repeatability under field conditions than the injury SOP maybe reflecting the multi-media training material that was used for BCS. To obtain reliable measures a training program is needed that provides a range of learning materials that takes into consideration different learning abilities.

Automated monitoring of lying time of dairy cows: on-farm sampling of animals based on individual differences

Vasseur, E., De Passillé, A.M. and Rushen, J., Agriculture and Agri-Food Canada, Agassiz BC, V0M 1A0, Canada; vasseur.elsa@gmail.com

The time that dairy cows spend lying down is an important measure of their welfare and activity loggers can be used to automatically monitor lying time on commercial farms. More and more dairy producers have activity loggers on their cows for estrus detection and these loggers could be used for this purpose. However, to obtain a representative measure of lying time in a herd, we need to know how to sample cows with respect to parity, stage of lactation and milk production. To determine how parity, stage of lactation and production level affected lying time, electronic data loggers recorded lying time for 10 d at 3 stages of lactation [S1=10-40, S2=100-140, S3=200-240 days in milk] of Holstein cows (mean±SD, milk production = 37.1±7.3 kg/d) with a parity ranging from 1 to 6 in tie-stalls (n=95). We calculated the total duration of lying time, the bout frequency and the mean duration of lying bouts. An increase in the stage of lactation increased the total duration of lying time (LS-mean±SE, S1: 10.8±0.3 h/d; S2: 11.4±0.3 h/d; S3: 13.0±0.4 h/d; generalized linear mixed model GLIMMIX P<0.0001) which was due to a reduced bout frequency (P<0.001) and increased mean duration of bouts (P<0.001). Parity and milk production did not affect total lying time at either S1 (n=82, P=0.2) or S3 (n=45, P=1.0) although higher milk production cows had a lower number of bouts at S2 (β=-0.014; SE(β)=0.007;P=0.04). At S2 (n=73), higher milk production cows had a lower time spent lying down (β=-6.582; SE(β)=2.597;P=0.01) but parity had no effect (P=0.03). At S1 and S3, higher parity cows had a lower frequency of lying bouts (β=-0.127; SE(β)=0.039;P=0.0017) but milk production had no effect (P<0.0001). There was a large variation between cows in lying time. The total duration of lying time was not correlated across cows between S1 and S2 (r=0.126; P=0.3) but was correlated between S2 and S3 (r=0.650; P<0.001). To estimate how the duration of the time sample affected the measures of lying time, subsets of data were created, consisting of 1 d to 9 d of observations. Measures based on each subset were correlated with measures based on the overall 10 d. The measures based on 4 d correlated highly with the measures based on all 10 days (r=0.95; P<0.001). Automated monitoring of lying time has potential as a measure of dairy cow welfare on commercial farms but cows differ greatly in lying time. It is necessary to sample cows based on their parity, stage of lactation and milk production level to obtain a representative measure for the herd. Targets for outcome-based measures need to take into account variation between animals.

Assessing the welfare of dairy calves: outcome-based measures of calf health versus input-based measures of the use of risky management practices

Vasseur, E.[1,2], Pellerin, D.[1], De Passillé, A.M.[2], Winckler, C.[3], Lensink, B.J.[4], Knierim, U.[5] and Rushen, J.[2], [1]Laval University, Quebec QC, G1K 7P4, Canada, [2]Agriculture and Agri-Food Canada, Agassiz BC, V0M1A0, Canada, [3]University of Natural Resources and Applied Life Sciences, Wien, A-1180, Austria, [4]Institut Supérieur d'Agriculture, Lille, 59046, France, [5]University of Kassel, Witzenhausen, 37213, Germany; vasseur.elsa@gmail.com

To examine whether outcome-based measures of calf health could be used to identify farms that use management practices that place calf health at risk, the mortality and morbidity of unweaned dairy calves and management practices that may impair calf health and welfare were surveyed on 115 farms in Canada (Quebec) and 60 farms in Central Europe (Austria and Germany). European farmers generally kept some records of calf mortality and morbidity but Quebec farmers usually only kept records of mortality at birth. Calf mortality was poorly estimated by producers in Quebec (Spearman correlation coefficient estimated vs. recorded data on mortality: $r=0.010$, $P>0.1$). Relative levels of morbidity on a farm could not be predicted from recorded levels of mortality (Spearman correlation coefficient; $r=0.142$, $P>0.1$). Health status was not necessarily associated with management practices generally recommended for health and welfare in either Quebec or Central Europe (Logistic regression, $P>0.1$). Quebec herds had higher calf mortality (25th percentile – median-75th percentile = 6.7/9.6/11.5) than Central Europe ones (0-5.4-20) and a greater use of management practices that may impair calf health and welfare were found in Quebec than in Central Europe ($\chi 2$ and Wilcoxon statistics, $P<0.1$); these were related to calving management and care of the newborn, colostrum management, calf feeding, weaning and calf housing. Inadequate recording of calf morbidity and mortality can be a problem in using farmers' health records to assess the level of calf health on a farm. The recorded mortality and morbidity do not necessarily show the extent that producers use management practices that pose a risk to calf health. Comprehensive assessment of dairy calf welfare requires the use of outcome based measures to assess the actual health and welfare status of calves combined with information on the use of management practices that place calf health and welfare at risk.

Inter-observer reliability of two systems for assessing body condition score of sows

Tucker, A.L.[1] and Lawlis, P.[2], [1]University of Guelph, Department of Animal and Poultry Science, Guelph, ON, N1G 2W1, Canada, [2]Ministry of Agriculture, Food and Rural Affairs, Animal Health & Welfare Branch, 401 Lakeview Drive, Woodstock ON, N1G 4Y2, Canada; penny.lawlis@ontario.ca

Body condition scoring (BCS) is the most common animal based measure for assessing welfare across species. It is important that auditors be trained to assess body condition to ensure that results are reliable. However, training and previous experience can impact on reliability of results for any animal based measure, including BCS. In addition, different types of scoring systems are available. The aim of this study was to investigate how different levels of experience influence inter-observer reliability across 2 different BCS systems. Fifty commercial cull sows were observed and scored by 1 expert and 17 graduate students in the 'Assessing Animal Welfare in Practice' course using both a 5-point scoring system (Cooperative Extension Service University of Kentucky; 1= thin to 5= fat) and a 3-point scoring system (Welfare Quality Assessment Protocol; 0=average, 1=thin, 2=fat). Training consisted of two, ½ hour classroom sessions with the expert 6 weeks and 2 weeks prior to data collection and included lectures and photo examples. Immediately prior to data collection at a livestock sale, students were able to compare body condition scores of 10 sows with the expert. Students rated themselves as having either no experience with BCS (N=12) or some experience (N=5). Data were entered into SAS (Version 9.1.3) and the influence of individual observer and level of experience were examined using a general linear model (PROC MIXED) for both BCS systems. Sow was included as a random effect in the model. Of the 50 sows observed, the expert scored 16% as 2 in the 5-point scale, 52% as 3, 30% as 4 and 10% as 5. Using the 3-point scale, 16% were scored as 1, 52% were 0 and 32% as 2. Overall, students differed significantly in how they scored sows using both the 5-point scale (P=0.012) and 3-point scale (P=0.051). Level of experience significantly influenced agreement with the 5-point scale between the expert and students having no experience (P=0.002) as well as students having no experience and some experience (P=0.003). Conversely, the 3-point scale detected only a trend for the expert to be different from the students with no experience (P=0.089). No other differences were detected between experience level using the 3-point scale. These results indicate that inter-observer reliability using the 5-point scale is more sensitive to training and experience than the 3-point scale and therefore more appropriate as an on-farm management tool versus a component for a welfare assessment program. Further work will examine the amount and type of training required to obtain consistent results using the 5-point scale.

Analysis of the cash euthanizer system in commercial production settings

Woods, J.A.[1,2], Hill, J.[3], Sadler, L.J.[1], Parsons, R.L.[1], Grandin, T.[4] and Millman, S.T.[1], [1]Iowa State University, Veterinary Diagnostic and Production Animal Medicine, 1600 South 16th Street, Ames, Iowa 50011, USA, [2]J. Woods Livestock Services, RR #1, Blackie, Alberta T0L 0J), Canada, [3]Innovative Livestock Solutions, RR 1, Blackie, Alberta T0L0J0, Canada, [4]Colorado State University, College Avenue, Fort Collins, Colorado, USA; jawoods@iastate.edu

The objective of this study was to assess the effectiveness of the Cash Euthanizer (CE) captive bolt gun (CBG) for single-step euthanasia. In the first phase of this study, 42 anesthetized pigs from 7 weight classes (2-3 kg, 7.5-10 kg, 15-20 kg, 30-40 kg, 100-120 kg, 200-250 kg, >300 kg) were euthanized using the CE which utilizes four head styles: a non-penetrating, short-length penetrating bolt, standard-length penetrating bolt and extended-length penetrating bolt. Four different power charges were utilized with the coordinating bolt and weight class. The placement of the CBG was based on American Association of Swine Veterinarians guidelines. The CE resulted in single step euthanasia for 38 of the pigs, with a trend for an effect of weight class for the secondary kill step (P=0.0951). All 4 pigs requiring a secondary step were over 200 kg, resulting in only 60% success rate for these top weight classes. Macroscopic evaluation of traumatic brain injury (TBI) was scored using a 3-point scale, and weight class effects were found for cerebral cortex (P=0.0068), thalamus (P<0.001) and cerebellum (P<0.001). In Phase II, 210 pigs representing the 7 weight classes were euthanized with the CE in a commercial production setting. Stockpersons at the production sites were responsible for the euthanasia of the pigs. 99.3% of the pigs under 120 kgs were effectively euthanized with the Cash Euthanizer. A second shot was required for 7 of the pigs over 200 kgs resulting in only a 77.7% success rate for the top weight classes. Casual observation suggests that stockperson error was associated with failure to euthanize with a single step, due to inadequate restraint or placement on the skull. In conclusion, the CE was found to be an effective and repeatable single-step method of euthanasia for pigs weighing less than 120 kgs. Further research is needed to identify changes in technique or technology for reliable euthanasia of mature sows and boars.

An assessment of the aggressive interactions of horses kept in large groups in a feedlot environment

Robertshaw, M.[1], Pajor, E.A.[2], Keeling, L.J.[3], Burwash, L.[4], Dewey, C.[1] and Haley, D.B.[1], [1]University of Guelph, 2538 Stewart Building, Guelph, ON, N1G 2W1, Canada, [2]University of Calgary, 3280 Hospital Drive NW, Calgary, AB, T2N 2Z6, Canada, [3]Swedish University of Agricultural Sciences, Box 7068, Klinikcentrum, Travvägen 10D, 750 07 Uppsala, Sweden, [4]Alberta Agriculture & Rural Development, 97 East Lake Ramp NE, Airdrie, AB, T4A 2G6, Canada; mrober11@uoguelph.ca

In western Canada, keeping horses outdoors in large groups (e.g. 100 to 200 animals) in pens with dirt flooring is a common method of management when feeding horses for meat production. We have found no published scientific studies documenting the behaviour or welfare of horses under these conditions. Because of their social evolutionary history and natural group sizes some have expressed concern that horses may not be well-suited for life in a feedlot environment. As a first step towards the development of measures for scoring horse welfare on farms, we documented the aggressive interactions of horses around 4 specific areas/resources within the pen (concentrate feeders, hay feeders, drinker, 'general' pen space). The 2 pens of horses observed contained (mean ± SD) 162.00±9.90 and 176.30±20.30 individuals, respectively, on observation days. Beginning in June for a period of 2.5 months, we recorded aggressive interactions by live observation on 2 days each week (a total of 240, 5-min intervals of observation/pen). Observations were made during a 2-h block of time each day, balanced across the following time periods: 07:00 to 09:00 h, 10:00 to 12:00 h, 13:00 to 15:00 h, and 16:00 to 18:00 h. During the 5-min observation interval, the total number of each of the following interactions was continuously recorded by one observer: CONTACTS: kicking, biting, displacement pushing, and rear-to-rear pushing, and THREATS: kick threat, bite threat, and ear pinning. The number of aggressive interactions was generally quite low with roughly 1 contact observed every 15 minutes. More contacts were recorded around the concentrate feeder ($P<0.00001$). Specifically, more biting and displacement pushing was observed there compared to the drinker ($P<0.01$) and the general pen space ($P<0.01$). There was also more rear-to-rear pushing at the concentrate feeder compared to the general pen space ($P<0.01$). Threats were more frequent than contacts, with roughly 1 threat observed every minute, however, the number of threats observed did not differ between the specific pen areas/resources.

Animal Welfare Judging & Assessment Competition (AWJAC): live & virtual assessments

Siegford, J.[1], Heleski, C.[1], Golab, G.[2], Millman, S.[3], Reynnells, R.[4] and Swanson, J.[1], [1]Michigan State University, 1290 Anthony Hall, East Lansing MI 48824, USA, [2]American Veterinary Medical Association, 1931 N Meacham Rd, Shaumburg IL 60173, USA, [3]Iowa State University, 2440 Lloyd Veterinary Medical Center, Ames IA 50011, USA, [4]United States Dept of Agriculture, 3466 Waterfront Centre, Washington DC 20024, USA; siegford@msu.edu

In 2001, Heleski, Zanella and Pajor proposed promoting animal welfare (AW) science to university students by coupling AW with traditional livestock judging. However, unlike traditional livestock judging, AWJAC requires students to examine the animals, as well as the conditions in which they are kept, to evaluate AW. In 2002, the 1st AWJAC had 4 undergraduate teams from 4 universities with 18 participants. In 2010, the 10th AWJAC had grown to 18 teams representing 9 universities with 78 participants in 3 divisions: undergraduate, graduate and veterinary students. Initially the AWJAC was conducted only with livestock species; now it encompasses production, companion, laboratory and exotic animals. In the AWJAC, students are asked to assess hypothetical, realistic scenarios containing performance, health, physiological and behavioral data as well as information about physical and social environments, human-animal interactions and management practices for two situations. Virtual scenarios, in a PowerPoint based format, are evaluated individually. Live scenarios are evaluated by teams of 3-5 students. Each team is allowed a short time to tour the facility and receives background information on management and veterinary practices that are not observable. After both types of assessments, students determine which facility has better welfare and present oral reasons regarding their rationale. Reasons are presented to a judge or panel of judges, in the case of the team assessment, with expertise in animal welfare science and specific knowledge of the species they are judging. During the team assessment, students may also be asked to recommend welfare-related improvements at the facility and answer judges' questions. Survey data collected after each competition show that >95% of participants believe the AWJAC is a valuable exercise, feel they have increased their knowledge about welfare science and would recommend the AWJAC to peers (n=345). In response to feedback, we expanded AWJAC to a 2-day format with a speaker program. This educational component has been extremely well received. While the assessment of various aspects of AW can be objective and quantifiable, judgment decisions of where on the continuum welfare is considered acceptable, preferred, or unacceptable often comes down to value-based choices. The AWJAC teaches students to integrate science-based knowledge with ethical values in an interdisciplinary approach to problem solving.

Provision of open water for farmed ducks: integration of methodologies and commercial application

Liste, G., Kirkden, R.D. and Broom, D.M., Centre for Animal Welfare and Anthrozoology, Department of Veterinary Medicine, University of Cambridge, Madingley Road, CB3 0ES, Cambridge, United Kingdom; gl318@cam.ac.uk

Different welfare assessment methods have been integrated in this project in order to design a commercially viable system which will provide farmed ducks with open water facilities. Farms rarely supply open water, citing management issues, health risks and effluent problems. The Council of Europe recommends that 'ducks should be able to dip their heads in water and spread it over their feathers' and the aim of this project was to find a practical way to implement this and incorporate it into the UK-RSPCA 'Freedom Food' label. The 'problem/solution' farm animal welfare investment model was applied to integrate the various elements (scientific knowledge, industry support and consumer pressure) necessary to achieve real change. Scientific knowledge. Firstly, the effects of different water sources on health and behaviour were investigated using small groups at commercial densities. Secondly a preference test was designed to assess duck behaviour and preference for different water depths under controlled, non-commercial conditions. Thirdly, a trial was designed to assess open water resources under commercial conditions, analysing health, production, water quality and water usage. Extrapolation of results from small groups to the commercial level proved challenging because health implications are different in large populations and behaviour varies with group size and dynamics. Industry support. In order to address producers' concerns and tackle management and environmental issues, the approach was to investigate the quality, decay rate and usage of water. An economic cost assessment was also planned considering data on production, labour, consumables, infrastructure, health, etc, from research and commercial outputs. It is easy to underestimate the importance of management issues before changes are in place (i.e. man hours needed to take care of open water resources). In addition, viability at the commercial level could be difficult to assess because buildings may need substantial changes to deal with open water (i.e. proper drainage) and provisional adaptations to test the new system could provide weak cost assessments. Consumer pressure. Freedom Food consumers regard welfare as their main concern, along with risks to their health and price. But if the research highlights poor industry practices, this could cause the consumers to refuse to buy the product. Moreover, they are paying a premium for this welfare label and may not accept a further price rise to improve standards. In summary, the final objective is to design a system to meet duck needs and prove viable in a commercial situation.

Improving utility of animal-based measures of welfare through amendment to audit protocol

Brown, A.F. and Twaissi, A., The Brooke, 30 Farringdon Street, London EC4A 4HH, United Kingdom; ashleigh@thebrooke.org

In response to emergent limitations in application and welfare significance of collected data, animal-based measures used by the Brooke for assessment of equine welfare were amended with the aim of improving utility. This paper focuses on two methodologies of change: adjustment to sensitivity of measures to abnormality, and subdivision of binary into multi-level scoring. Data from 350 working equids in Petra, Jordan collected via the original and new measures for body lesions, hoof horn quality and eye abnormalities were compared to identify how the amendments altered the information obtained. The measure of body lesions was amended by increasing sensitivity to abnormality. With the original measure, data for lesions on the head/ears, girth/belly and withers/spine areas showed prevalence of 2%, 5% and 1% respectively. With the new measure, prevalence in the same areas was 8%, 11% and 5% respectively. Therefore 74% of head/ears lesions, 55% of girth/belly lesions, and 81% of withers/spine lesions captured by the amended measure were excluded by the original. The measure of hoof horn quality was amended by decreasing sensitivity to abnormality. Data from the original measure showed 100% prevalence of abnormality in one or more hooves compared to 42% with the new measure. When hoof horn quality was compared in population sub-groups of 128 horses and 136 donkeys/mules, the new measure showed 50% abnormality prevalence in horses and 32% in donkeys/mules. With the original measure both groups scored 100% abnormality. The measure for eye abnormalities was amended by altering from binary to multi-level scoring and decreasing sensitivity to abnormality. With the original measure results showed 100% prevalence of eye abnormality; with the new measure, 13% prevalence of moderate abnormality and 14% prevalence of severe abnormality were detected. Comparison between the same sub-groups revealed that horses had 18% moderate and 9% severe eye abnormality, and donkeys 11% moderate and 24% severe abnormality. With the original measure both groups scored 100% abnormality. Data collection with the new measures was not more time-consuming or intrusive to the animal, yet far more informative, thus increasing efficiency of invested resources. There was improved capacity to detect pertinent differences between groups of animals, and consequently increased sensitivity to change over time – the former essential to identify animals in poorest welfare, the latter important for monitoring of welfare change. This work demonstrates the benefits of context-appropriate adaptability in assessment protocol and has potential for application to other environments where animal-based assessment of welfare is a constituent of management or monitoring systems.

Relationship between piglets skin temperature measured by infrared thermography and environmental temperature in a vehicle in transit: a preliminary study

Nanni Costa, L.[1], Redaelli, V.[2], Magnani, D.[1], Cafazzo, S.[1], Amadori, M.[3], Razzuoli, E.[3], Verga, M.[2] and Luzi, F.[2], [1]Univ. of Bologna, DIPROVAL, via G. Fanin 46, 40127 Bologna, Italy, [2]Univ. of Milano, DSA, via G. Celoria 10, 20133 Milano, Italy, [3]IZLER, via Bianchi 7/9, 25124 Brescia, Italy; veronica.redaelli@unimi.it

The thermal state of pigs during transport is difficult to assess because it depends on the mixed of production of heat and humidity, of external environmental conditions and of the vehicle ventilation. The measure of body temperature has a particular relevance for the assessment of pig welfare during transport, but this parameter is difficult to detect in transit. In the last two decades, infrared thermography (IRT) has been already applied on pig in order to detect their superficial body temperature, but not during the journeys. Infrared thermography does not require any contact and is therefore a completely non-invasive technique that allows to record temperature of hardly approaching or moving subjects. The purpose of this study was to detect, continuously, a thermometric profile of piglets skin during a long journey and to evaluate its relationship with the temperature inside the vehicle. Two journeys of 14 hours were carried out on July and September 2009 respectively and skin temperature variations were measured by infrared thermography on a total of 12 piglets, six for each trip. Average weight of piglets transported was 10.88±1.97 kg in the first trip and 8.08±0.74 kg in the second one. A thermocamera Avio TVS 500 was placed in front of and above the pen, in the first and in the second journey respectively. Environmental temperature and humidity inside the compartment housing piglets were continuously recorded. Relationship between piglets skin temperature max value and environmental temperature in the vehicle during the two transports was examined by regression analysis. During the second trip, camera location above the pen led to higher values of skin temperature reducing also the variability of thermal measurements, compared to the first one. In both journeys relationship between skin temperature and environmental temperature inside vehicle was linear ($P<0.001$). The R^2 value was equal to 0.44 and 0.57 in July and in September, respectively. Thus, in the range of temperatures recorded during these transports, an increase of one degree in temperature inside the vehicle matched an increased of 0.2 °C in piglets skin temperature. In our knowledge, this is the first study on using infrared thermography to measure continuously changes in skin temperature on pigs during transport. This technique, coupled with deep temperature recording systems, will help to better understand the adaptive efforts of piglets to extreme environmental conditions experienced during transport.

Infrared thermography: a non invasive technique to asses metabolic activity in horses

Valle, E.[1], Redaelli, V.[2], Papa, M.[2], Bergero, D.[1] and Luzi, F.[2], [1]International Center of the Horse La venaria Reale, Cascina Rubbianetta Druento, 10040 Druento, Italy, [2]Univ of Milano, DSA, Via G. Celoria, 10, 20133 Milano, Italy; fabio.luzi@unimi.it

To understand the normal variations in equine thermal patterns using Infrared thermography (IRT) is crucial for both clinical cases and welfare procedure. The purpose of this study was to investigate temperature (T) changes in specific body regions (coronary band (Cb) used as indicator of abuse of substances to influence performance and eyes (E) as indicator of core body T) using IRT according to the physiological changes after feeding. Six healthy warmblood horses (geldings) (5 to 10 years old) (medium BW 555±73 kg) were included in the study and kept at International Center of Horse. The animals were fed with maintenance ration composed by 80% first cut meadow hay and 20% of oat. The feeding plan was divided in three meal a day: the same amount of oat was given daily at 08:00 am and 01:00 pm and 06:00 pm; hay at 08:00 am and 06:00 pm. IRT images and rectal temperatures (RT) were recorded before the meal (morning meal (mm), noon (nm) and evening (em) and after 10 minutes (IR10), 60 (IR60), 180 (IR180) and 240 (IR240) after each meal and every two hours during the night. IRT images were taken from the dorsal view of forelimb to monitor T of Cb and later view for eyes E surface. The max T within the area of interest was recorded. After the normality was checked the mean changes Cb and E T with time was compared using repeated measure ANOVA; Bonferroni post hoc test was used to identify differences ($P<0.05$); Pearson correlation was also performed between RT, Cb and E. RT increased from a value of 37.5±0.24 mm to the value records at IR60nm (37.9±0.19 $P<0.01$), reaching the max value 37.9±0.19 for IR240em ($P<0.01$) and returns to basal values at 2:00 am. No variation was identified comparing the T recorded before and after meal. T of E increased gradually from 35.4±0.38 mm to the values records at IR60nm (36.5±0.34 $P<0.001$). Maximum value was at IR60em (36.9±0.32 $P<0.001$) and returned to basal value at 4:00 am. No variation was identified comparing the T recorded before each meal and after. Cb T increased immediately from a value of 32.7±0.62 mm to IR10mm (33.8±0.37 $P<0.05$); this increase persisted during day and night reaching max value at IR10em (35.1± 0.65). RT was correlated with Cb ($P<0.01$) and E ($P<0.01$). RT variations recorded were less than 0.4 °C; E variations recorded were less than 1.5 °C; CB variations recorded were less than 2.4 °C. In the last case the values recorded seems to be influenced not only by the circadian rhythm of body T but also other metabolic activity such as the meal. Identify the standards of normality could help the practitioner to identify horse's with abnormalities or hypersensitizing substances abuse.

Are modified bell drinkers a suitable water source in the Pekin duck production, concerning animal health and behavior?

Bergmann, S.[1], Heyn, E.[1], Schweizer, C.[1], Damme, K.[2], Zapf, K.[2] and Erhard, M.H.[1], [1]Chair of Animal Welfare, Ethology, Animal Hygiene and Animal Housing, Department of Veterinary Sciences, Faculty of Veterinary Medicine, LMU Munich, Veterinärstr. 13/R, 80539 Munich, Germany, [2]Bavarian State Research Center for Agriculture, Specialization in Poultry Management, Mainbernheimerstr. 101, 97318 Kitzingen, Germany; s.bergmann@lmu.de

Pekin ducks for meat production in Germany are held without access to open water, mainly out of practicable and hygienic reasons. Nipple drinkers are the commonly utilized drinking system for these water birds. The aim of this study was to analyze the feasibility of modified bell drinkers as a species-appropriate water supply for Pekin ducks under farm condition, regarding animal health and behavior. In collaboration with three duck farms (8,500 to 13,500 ducks/unit) a system of modified bell drinkers (Big Dutchman, Vechta, Germany) was installed on the incline side of each duck farm building. This drinking system was offered to the ducks during a time period of six hours daily, starting at an age of 25 days until slaughter. Five to eight mast periods per trial were examined on each farm, while test trials (plus modified bell drinkers) and control trials (exclusively nipple drinkers) took place alternately. During the mast periods the farms were visited twice in a defined time frame. Per visit 100 randomized ducks underwent an animal health examination and video observation (six to twelve cameras/ farm) was started. Compared to control trials the Pekin ducks used significantly ($P<0.01$) more water (0.473 l/duck vs. 0.550±0.007 l/duck) for drinking, to dunk their heads, scoop up water and to clean their feather coat during the test trials. The drinking activity (percentage of ducks being busy with drinkers) while having access to the bell drinkers (test trials, up to 90%), was significantly ($P<0.001$) higher comparing control trials (0-45%). In total 5.5-22.0% of the examined ducks (8,200) showed unilateral and 0.7-3.7% bilateral congestion of the nose openings. The eye surroundings were severely soiled in 0.4-1.9%. Concerning both parameters, ducks with access to the modified bell drinkers had a significantly ($P<0.001$) lower percentage in comparison. The average mortality was located between 2.4% (test trial) and 2.7% (control trial). The results show that modified bell drinkers are a suitable water source for ducks under farm condition. The modified bell drinkers allow the ducks an efficient fulfillment of behavioral patters. Out of economic reasons it is important to reduce the access to the modified bell drinkers for a certain time span, to lower the costs for water, labor and additional litter. A six hour access seem suitable.

Evaluation of poultry welfare on Lithuanian farms

Ribikauskas, V. and Skurdeniene, I., Lithuanian University of Health Sciences, Institute of Animal Science, Zebenkos 12, LT-82317 Baisogala, Radviliskis Distr., Lithuania; vytautas@lgi.lt

Research into poultry welfare was conducted in 30 Lithuanian poultry farms. The aim of the work was to develop and check poultry welfare evaluation questionnaire projects in respect of six poultry groups kept for farming purposes. The following criteria were evaluated: (1) possession of relevant qualifications by the staff maintaining poultry farms; (2) regular inspection of poultry; (3) health of poultry; (4) freedom of poultry movement; (5) buildings and premises; (6) feeding, feeding quality, watering and water quality; (7) evidence of technological mutilations; (8) breeding and reproduction. This summary presents the results of the evaluation of buildings and premises. Optimal room temperature was recorded on 57.1% of laying hen farms, 62.5% of chicken broiler and replacement pullet farms, and 71% of turkey farms. During the cold season, a room temperature of 4.2-9.2 °C was recorded in 66.6% of duck and geese farms, which is close to outdoor temperature. The dust concentration on the evaluated poultry farms ranged from 0.02 to 30 mg/m^3; in broiler accommodations from 0.03 to 16.4 mg/m^3, turkey accommodations from 0.12 to 0.46 mg/m^3, and duck and goose accommodations from 0.03 to 0.10 mg/m^3. The ammonia concentration did not exceed the 10 ppm limit in 85.7% of hen keeping rooms. The ammonia concentration exceeded the 40 ppm limit in 16.7% of the evaluated accommodations used to keep broilers and replacement pullets, and in 33.3% of the evaluated poultry farms it exceeded the 20 ppm limit. In 11% of turkey rooms the ammonia concentration exceeded 10 ppm, and in 16.7% of duck and goose rooms it exceeded 20 ppm. On 18.2% of laying hen farms the CO_2 concentration was higher than the maximum allowable value of 2,000 ppm, i.e., the ventilation system did not operate at adequate efficiency. Total bacteria count in the air of laying hen rooms ranged from 4.3 to 3,045.2 thousand per cubic meter; in broiler rooms 2.0-360.0, turkey rooms – 26-461, duck and goose rooms – 0.3 to 461. Artificial lighting intensity in the cage zones on laying hen farms was 4.0-25.0 lx. Where use was made of natural lighting (conforming to the daily rhythm), its intensity was 20 to 60 lx. Artificial lighting in broiler accommodations was 9.0 to 45.0 lx, in turkey accommodations 10.0 to 45.0. 40% of the evaluated broiler farms used lighting regime programmes. 54% of farms produce chicken broilers without providing the darkness period of rest (that is, the photoperiod was 24 h per day). 28.6% of turkey farms use natural lighting only, and 28.6% use red or blue light bulbs to prevent cannibalism. Based on the research results, a simplified evaluation of poultry welfare in questionnaire form has been developed and adapted for provision of sufficient comparative information.

Health characteristics of fattening pigs reared at pasture compared to 5 indoor housing systems

Tozawa, A.[1], Inamoto, T.[2] and Sato, S.[1], [1]Tohoku University, Graduate School of Agricultural Science, 232-3, Yomogita, Naruko-Onsen, Osaki, Miyagi 989-6711, Japan, [2]Akita Prefectural University, Faculty of Bioresource Sciences, 241-438 Kaidobata-Nishi Nakano, Shimoshinjo, Akita 010-0195, Japan; akitsu-t@bios.tohoku.ac.jp

Pigs reared outside are exposed to more pathogens and unfavourable weather, potentially compromising health and productivity. This study compared the health of fattening crossbred pigs reared outside to that of pigs reared in 5 indoor fattening systems. Health of pigs turned out to pasture from 109 days of age (Farm A) was compared to that of pigs sampled from 5 different fattening systems (Farm B-F). In all cases (except Farm C), pigs were sampled from a single pen. Farm A, 9 pigs (LWB) given 840 m^2 of pasture and 5.56 m^2/pig of indoor area; Farm B, 9 pigs (Landrace) in a concrete floored indoor pen of 2.53 m^2/pig with an outside concrete paddock of equal size; Farm C, 10 pigs (LWB) from indoor pens (9 pigs/pen, 1.11 m^2/pig) with a part solid and part slatted floor; Farm D, 10 pigs (LWD (SPF)) from an indoor pen (20 pigs/pen, 0.68 m^2/pig) with a fully slatted floor; Farm E, 10 pigs (LWD (SPF)) from a large fully slatted pen (350 pigs/pen, 0.74 m^2/pig); Farm F, 10 pigs (LWD (SPF)) from a deep litter pen (350 pigs/house,1.71 m^2/pig). Pens in Farms D-F were windowless. Serum, saliva, and feces were collected before slaughter for immunogloblin A (IgA) quantification. Pigs at Farm A were reared at Farm C until being moved outdoors. Incidence of pneumonia from M. hyopneumoniae (MPS score) of Farm A and C pigs were compared at slaughter. Intestinal flora from feces were compared using 15 culture media during the indoor and outdoor phases at Farm A. Average daily gain (ADG) was also measured. Farm A had the lowest mean serum IgA level (1.77E+06 ng/ml; significantly lower than Farms B, C, and E; One-way ANOVA, $P<0.001$). Mean salivary IgA was second lowest on Farm A (7.15E+04 ng/ml; significantly lower than Farm D; One-way ANOVA, $P<0.001$), whilst a moderate mean fecal IgA level was seen on Farm A (30.22 ng/mg; significantly higher than Farm E; One-way ANOVA, $P=0.005$). ADG to slaughter was lowest at Farm A (0.568 kg/day) as a result of the early period of outdoor rearing when pigs had not acclimatized to the hot summer conditions, whilst ADG subsequently recovered in the later stage of outdoor rearing (Two-way ANOVA, $P<0.001$). Mean MPS score at Farm A (10.91%) tended to be lower than Farm C (17.73%; Mann-Whitney test, $P=0.057$). Intestinal flora were reduced in 10 culture media at Farm A after outdoor compared to indoor rearing (significant in 4 culture media; Paired Student's t-test, $P<0.05$). Despite the risk of pathogen exposure, fattening pigs reared at pasture have lower IgA levels and a lower number of intestinal flora.

The implementation of animal welfare in GLOBAL GAP for livestock producers and the impact on auditing – a practical, global perspective

Aumueller, R.K. and Coetzer, E., GLOBALG.A.P, Standards Management Livestock and Feed, Spichernstr. 55, 50672 Colgne, Germany; aumueller@globalgap.org

Certification of primary producers at the farm level according to GLOBALG.A.P´s (GG) Integrated Farm Assurance standard (IFA) is a holistic approach that incorporates food safety, environmental, social and animal welfare criteria. The integration of animal welfare-related criteria in IFA for livestock producers is based on applicable legislation, as well as demands of GG members and the market. Demands from retail members and society to include additional, voluntary and easily-auditable animal welfare criteria that go above legal requirements are growing. Animal welfare criteria must be discussed with producers and certification bodies (CB) to validate auditability, practicality and economy. Feasibility and acceptance of criteria must be demonstrated in pilot audits. GG IFA Version 4, released March 2011 was developed through a revision process by sector committees having a 50:50 representation of producers and retailers. Relevant modules of IFA for livestock producers are All Farm Base (AF), Livestock Base (LB), Ruminant Base (RB), and species-specific modules: Cattle and Sheep (CS), Dairy (DY), Calf Young Beef (CYB), Pigs (PG), Poultry (PY), Turkey (TY). Under IFA livestock, housings and facilities, health and fitness of the livestock, procedures for loading and unloading livestock must be reviewed via 3rd party audits from approved CBs. For approval the CB's auditors must undergo initial GG training with periodic follow-up courses. In LB, 29 out of 107 (26.7%) control points are related to animal welfare. For the species-specific modules, these figures are: CS: 22/30 (73.3%); DY: 24/58 (41.4%); PG: 44/80 (55.0%), PY: 88/159 (55.4%) and TY: 58/101 (57.4%). Currently, GG is developing voluntary customized add-on modules that outline further criteria and requirements to complement the existing mandatory modules. The voluntary modules are designed as species-specific add-on tools to ensure higher levels of animal welfare in livestock production. Development of these voluntary add-on modules started with pigs and poultry, based on the expectation that these two species will play the most important role in the global supply of animal protein in the future. Livestock transport has aroused much debate over the past years. Based on EU regulations, the transport standard provided by GG takes into account differences in transport regulations across countries. Standard Benchmarking to reduce 'auditing tourism' on livestock farms is strongly supported. Producers will benefit from the 'one-stop audit at the farm gate' advantage through a reduction of audit duplication and auditing costs, as well as the international recognition of the GG standard.

Implementing the Global Animal Partnership (GAP) 5-step animal welfare program: a retailer's perspective

Malleau, A.E. and Flower, F.C., Whole Foods Market, 550 Bowie St, Austin, TX, 78703, USA; Anne.Malleau@wholefoods.com

Despite extensive work on on-farm welfare measures and their validation, little work has reported on the implications of welfare programs to retailers. In Jan 2010, Whole Foods Market (WFM) announced all pork, beef and chicken vendors were to be certified to the GAP 5-Step Animal Welfare Program. With 60 meat vendors and more than 1,200 farms, implementation was challenging, especially as every farm needed to be third-party audited by Jan 2011. To achieve nationwide implementation, we had to solve challenges at the farm level, demonstrate chain of custody from farm to store, and execute the program at the retail level. Major challenges at the farm level differed between species. To tackle the various challenges, WFM partnered with our suppliers to offer pre- and post-audit support programs. The pre-audit program included a variety of preparation tools, resources, farm manuals, and record sheets focusing on animal-based measures. WFM also offered pre-site visits, help completing applications, and 55+ vendor-specific workshops. Our post-audit program helped farmers with corrective action plans and supporting evidence when necessary. To make in-store animal welfare label claims, WFM developed a chain of custody program. The program requires documentation to travel with each group of animals and once they are processed, documentation is required on each box of product before going to distribution centers. When product arrives at the stores, there are specific protocols to ensure meat in the coolers, service and fresh cases are properly labeled with the supplier's step-rating. Execution of the retail program involved a multifaceted approach, based on individual assessments of a supplier's situation and system. Examples included: cost sharing of audit fees; bids; formal commitments; volume commitments; and carcass utilization. Additional considerations at store level included clearing inventory before the launch, training team members, creating a marketing campaign, and developing strategies to 'tell the story' to customers. Although this was the first time WFM required 3rd party animal welfare verification of our entire farm community, the implementation process was extremely successful (99.4% pass rate). To date, 1,239 farms are currently certified from the U.S., Canada, and Australia supplying Step-rated meat ranging from Step 1 to Step 5. WFM took a hands-on approach, supporting suppliers and acting as a catalyst to make the Program happen. Roll-out of an animal welfare program takes 100% commitment and collaboration at every level (farm, supplier, audit company, retailer, GAP) to achieve success.

Robot milking does not seem to affect whether or not cows feel secure among humans

Andreasen, S.N. and Forkman, B., University of Copenhagen, Department of Large Animal Sciences, Groennegaardsvej 8, 1870 Frederiksberg C, Denmark; sinen@life.ku.dk

In Denmark, as in many other countries, robot milking (AMS) has made its entry into the dairy farms. The use of robots has supposedly decreased the interaction between the dairy cows and the farmer. A possible negative consequence of AMS on animal welfare could be a worsened Human-Animal relationship, which could result in animals that are fearful of humans. In this study the avoidance distance for cows from 24 farms with an ordinary milking parlor (mean number of cows tested on each farm = 67 (49-87), in all 1,617 cows) was compared to 16 farms with AMS (mean number of cows tested on each farm = 72 (58-83), in all 1,155 cows). The average number of animals on milking parlor farms was 196 animals and the average number of animals on farms with AMS was 214 animals. In all 40 farms with Danish Holstein-Frisian cattle were included in the study. The avoidance distance of the cows was measured according to the European Welfare Quality® protocol. The test started approximately 15 minutes after morning feeding and was carried out at the feeding table. The test person was unfamiliar to the cows and the sample size set by the European Welfare Quality® protocol was used. For each farm the average avoidance distance in centimeters was calculated and a t-test of the farm means was performed. The result showed that there was no significant difference in the avoidance distance between cows milked in ordinary parlors and cows milked in AMS (P=0.57, t-value = 0.57, N=40). To investigate if the number of animals had an effect on the avoidance distance a correlation between the avoidance distance and the number of animals on the farms was calculated, this was however, not significant (R_s= -0.0052, P=0.97). The results show that the postulated decreased interaction between cows and handlers when using AMS does not affect the Human-Animal relationship when the avoidance distance test is used as a measure. Earlier investigations have indicated that Human-Animal relationship may affect the production results of dairy cattle. On 31 of the 40 farms data concerning milk yield was available. On basis of these data the results of the earlier investigations are not supported in the current study, neither for the cows in AMS (avoidance distance/milk yield – r^2=0.1; P=0.31) nor for the conventionally milked cows (avoidance distance/milk yield – r^2=0.06; P=0.32).

Assessment of pig welfare in different housing systems

Juskiene, V., Juska, R. and Ribikauskas, V., Lithuanian University of Health Science, Institute of Animal Science, R. Zebenkos 12, Baisogala, LT-82317, Lithuania; violeta@lgi.lt

In recent years, the public focus on farm animal welfare has increased expressing concern for the effects of intensive livestock production systems on animal welfare. Livestock welfare has become a production parameter like productivity or product quality. Improvement of the animals' housing conditions could positively affect their welfare, and consequently their health and productivity. Therefore, this study was designed to evaluate different growing technologies applied on the farms of Lithuania in relation to pig health and welfare. The studies of pig health and welfare were carried out during farm visits on the basis of the ANI 35L system. Animal welfare was evaluated based on following indexes: freedom of movement, social contacts, floor type, microenvironment and pig care – in all 43 direct and indirect indexes. Various technological pig groups were evaluated: pregnant and non-pregnant sows, piglets under 90 days of age (under 30 kg) fattening pigs (under 110 kg). According to the housing system, the sow group was divided into those kept littered and unlittered. Then housing conditions for piglets under 90 days of age were evaluated with regard to their raising indoors under controlled microenvironment and in ordinary pig houses on scarce litter. Besides, deep litter housing of fattening pigs and their housing on ceramic tiles was evaluated with respect to pig welfare requirements. The study indicated that bedding of sows that were kept on concrete- ceramic tile floor and sow keeping without any bedding had no significant effect on their welfare. A significantly better (62.9%, P=0.001) housing conditions were determined for the piglets under 90 days of age kept in ordinary pig houses with uncontrolled microenvironment. Fattening pigs housed on deep litter also had more suitable conditions than those housed on unlittered concrete-ceramic tiles and the average score was by 6.2 points (P=0.021) higher for the fattening pigs housed on deep litter. The results of study demonstrate that for the improvement of pig welfare is advisable to use some bedding material.

Preference and lying time of beef steers housed on rubber-covered slats vs. concrete slats

Plaster, S., Jobsis, C.T. and Schaefer, D.M., University of Wisconsin – Madison, Department of Animal Sciences, 1675 Observatory Drive, Madison WI 53706, USA; plaster@wisc.edu

Feedlot cattle in the Midwest are often housed on raised concrete slats. There are anecdotal claims that covering the slats with rubber matting improves cattle comfort. Our objective was to determine if cattle have a preference for rubber surface over concrete surface and if flooring type has an impact on time spent lying. The project was divided into two phases. During the first phase, two pens (12.6 m × 3.6 m) were used; half of the floor of each pen was concrete slats (CS) and the other half was rubber covered concrete slats (RS). Sixteen beef steers were weighed and 8 randomly assigned to each pen. Cattle were fed a corn and corn silage-based diet, and consumption for the pen of 8 animals was recorded daily. Two cameras were positioned to capture photographs every 5 min, 24 h per day. After 2 weeks, the second phase began. A gate was closed in the center of each pen thus forming 4 pens (6.3 m × 3.6 m); 2 pens with concrete slatted flooring and 2 pens with rubber-covered slatted flooring. Cattle were weighed on day 14 and were randomly assigned to a pen. Cattle were housed in pens of 4 animals on either concrete or rubber surfaces for an additional 14 d. The two cameras continued to capture a photograph every 5 min for the next 14 d. Feed consumption was monitored for each pen of 4 animals during phase two of the project. The phase one photographs were viewed and the number of cattle on the concrete or on the rubber was recorded. Location (CS or RS) was converted to time spent on the two types of flooring. Statistical t test analysis of phase one cattle location data revealed a strong preference for rubber over concrete slats ($P<2.2^{e-16}$). Body position, standing or lying, was also measured. These data were calculated into minutes spent lying vs. standing for each pen of animals. The animals displayed a strong preference to the rubber covered slats when lying down with over 92% of their lying down time spent on the RS. The phase two photographs were viewed and the position (lying or standing) was recorded for each animal in each pen. Analysis of phase two data revealed that the amount of time spent lying on RS vs. CS was not different. Feed consumption during phase 2 was significantly higher ($P=2.86^{e-08}$) for animals housed on RS. Body weights did not differ between groups.

Analysis about the accurate point of encounter during stunning of cattle with the captive-bolt pistol

Bergmann, S., Schweizer, C., Heyn, E., Kohlen, S. and Erhard, M.H., Chair of Animal Welfare, Ethology, Animal Hygiene and Animal Housing, Department of Veterinary Sciences, Faculty of Veterinary Medicine, LMU Munich, Veterinärstr. 13/R, 80539 Munich, Germany; s.bergmann@lmu.de

The stunning of animals for slaughter should be carried out fast, reliable and with avoidance of any pain or suffering. The captive-bolt stunning is at present the most common practiced method for stunning cattle in the European Union. The aim of this study was to investigate, if the recommended point of encounter for stunning cattle with a captive-bolt pistol, fits even special cattle breeds so that the bolt will penetrate the brain in direction of the brain stem and caudal parts of the hemispheres. In common practice it could be seen that there are anatomical characteristics, that could lead, even with a correct point of encounter, to not or insufficient stunned individuals. In this study 1,027 adult cattle (516 bulls, 125 heifers and 386 cows) out of 26 different breeds were examined during routine slaughter in an EU approved slaughterhouse in Germany. After the regular slaughter process, the heads were directly measured on seven defined surface areas and individually marked at the in Germany recommended point of encounter: 'cross-over point of the connecting line from the center of the horn to the center of the opposing eye'. Then each head was parted median. The previous marked recommended point of encounter and the, when existent, under the anatomical aspects ideal point of encounter were opposed to each other and the separation distance between the both was measured. In 54.8% of the examined animals there was no match between the recommended point and the anatomical ideal point, which means that there occurred a deviation from these two points. In 1027 examined heads exclusively positive deviations could be seen, which means that the anatomical ideal point would lie above the recommended point. Bulls and heifers have a 30.7% and 16.2% higher risk for a positive deviation than cows. Cows therefore would have been hit significantly more often (χ^2-= 29.7, df=3, P<0.001) in the anatomical ideal point (56.1% of the cases). The arithmetic mean of the positive deviation lies at 2.72 cm (± 0.0013). The breeds Charolais (2.82±0.011 cm), German Simmental (2.87±0.004 cm) and crossbreeds show a significant higher risk (P<0.001) of a positive deviation compared to Holstein Friesians (black: 2.44±0.011 cm and red: 2.59±0.048). The results show, that the risk for a positive deviation of the point of encounter, using bolt pistols for stunning cattle, is influenced significantly by breed, sex and weight. It is therefore required to heighten the at present recommended point for cattle.

Can pigs discriminate using social categories?

Jones, S.[1], Burman, O.[2] and Mendl, M.[1], [1]University of Bristol, Animal Welfare and Behaviour Group, Langford, BS40 5DU, United Kingdom, [2]University of Lincoln, Biological Sciences, Lincoln, LN2 2LG, United Kingdom; s.m.jones@bristol.ac.uk

Although positive social relationships can enrich the lives of animals, aggression, intimidation and socially induced stress can cause serious welfare problems. It is therefore important to try to understand how animals remember other individuals, and past social encounters. One possible way in which animals discriminate between other individuals is by developing a 'familiarity'/'group membership' memory, or categorisation. We studied this using a Y-maze task in which test pigs were each trained to discriminate between two stimulus pigs (one group-mate (GM) and one non-group mate (NGM)) situated at the end of each maze arm. Twenty-one female Large White/Landrace pigs were housed in three groups of seven. Test subjects (n=2 from each group) were trained to approach one stimulus animal ('positive' stimulus (GM for 50% of test pigs)) by rewarding approach with food, whilst movement towards the other animal (`negative` stimulus) was unrewarded. Subjects received ten tests per session, one session per day and stimuli were pseudo-randomly switched between the two arms of the maze. Subjects were tested with three different stimulus pairs, progressing from one pair to the next once the test animal reached a criterion of ≥8/10 correct responses in two consecutive sessions (one tail binomial, P=0.003). Five pigs reached criterion using all three different stimulus pairs and sessions to criterion decreased with each subsequent pairing (Friedman(2)=8.316: pair sets 1-3, P=0.008). We used reversal trials, in which the reward contingency was reversed (positive stimulus became the negative stimulus), to determine if the subjects had actually learnt the categorical discrimination, or whether they had simply learnt to rapidly identify the rewarded exemplar in each new pair without developing the appropriate cognitive categories. The mean number of errors in the first 20 trials significantly increased during reversals (Friedman(2)=13.776: pair sets 1-3, P<0.001; Wilcoxon: pair set 3 vs.reversal 4: Z=-2.032, P=0.031) suggesting that the pigs had indeed learnt to discriminate between group-mates/non-group-mates. This design can now be adapted to determine how long pigs can remember conspecifics allowing animals to be removed and re-introduced to groups whilst minimising social stress. The research therefore offers insight into how animals may categorise conspecifics, furthers our understanding of the social structure of group-housed animals, and suggests ways in which management may be altered to minimise aggressive interactions and associated bodily damage and stress. The results of this study therefore have potential implications for both animal welfare and agriculture.

The 'auntie effect': presence of an older companion reduces calf responses to weaning

De Paula Vieira, A., Von Keyserlingk, M.A.G. and Weary, D.M., University of British Columbia, Animal Welfare Program, 2357 Main Mall, Room 190, V6T1Z4, Canada; apvieirabr@yahoo.com.br

In commercial dairy production calves are typically separated from the dam at a young age. This practice may interfere with developmental processes mediated by social interactions that occur between the calf and older social partners. These social influences may be especially important in the development of feeding behavior, and social learning may help smooth the transition from milk to solid feed and thus minimize weight losses at weaning. The aim of this study was to test the prediction that calves kept the company of an older, weaned calf would show earlier intakes of solid feed before weaning, a lower growth check at weaning and better growth rates after weaning, compared to calves housed with calves of their own age. Forty-five dairy calves were separated from the dam and housed in pens composed of either groups of 3 unweaned calves, or groups of 2 unweaned calves and an older weaned calf. The group pens were equipped with automatic milk, water, starter and hay feeders. Weaning from milk was by gradual reduction of milk volume over 5 d, from d 36 to d 40. During the pre-weaning period (d 1 to d 35) the number and duration of visits to the hay feeder was higher for calves housed with an older companion, and calves in this treatment consumed more hay (58 vs. 26±20 g/d). Starter intakes did not differ during this period, but the number of visits to the starter feeder was higher for calves housed with an older companion (156.2 vs. 99.8±54 visits). During the weaning and post-weaning periods, calves housed with an older companion consumed more starter (1.54 vs. 1.11±0.13 kg/d) and made fewer unrewarded visits to the milk feeder (7.8 vs. 11.7±2.1 visits/d). Over the entire experimental period, calves housed with an older companion gained more weight (pre-weaning period: 0.89 vs. 0.76±0.04 kg/d; weaning and post-weaning periods: 1.25 vs. 1.04±0.07 kg/d). We conclude that housing young calves with an older companion stimulates feeding behaviour before and after weaning, improving growth rates. Higher intakes of solids may also reduce calf hunger around weaning, explaining the reduced number of visits to the milk feeder by calves housed with older companions.

Structural MRI of piglet brain development from postnatal day 1 through 28

Nordquist, R.E.[1,2], Zeinstra, E.C.[1], Van Dijk, L.[1], Van Der Marel, K.[3], Van Der Toorn, A.[3], Van Der Staay, F.J.[1,2] and Dijkhuizen, R.M.[3], [1]Utrecht University, Emotion & Cognition Res Prog, Dept Farm Animal Health, Yalelaan 7, 3584CL Utrecht, Netherlands, [2]Utrecht University, Rudolf Magnus Institute for Neuroscience, Universiteitsweg 100, 3584 CG Utrecht, Netherlands, [3]Utrecht University, Image Sciences Institute, Universiteitsweg 100, 3584 CG Utrecht, Netherlands; r.e.nordquist1@uu.nl

Pig husbandry practices have caused a sharp increase in the number of piglets produced per litter, which correlates with increased numbers of low birth weight piglets. This and other early life adverse events may have profound consequences for brain development and for welfare. Measures of brain morphology and physiology have been suggested to reflect welfare states in animals. However, in order to determine the extent of adverse effect, normal growth curves for brain development are needed. To address this need for baseline values, brains from 15 female piglets ([(Terra × Finish landrace) × Duroc]; 1,130-1,930 g birth weight) were collected, followed by postmortem structural MRI at 1, 14, or 28 days postnatally (n=5 per time point). Brain volume was measured with Medical Image Processing, Analysis and Visualization (MIPAV) software, hand corrected when needed. Total brain volume in cm3 at 1 day was (mean ± SEM) 39.88±2.20; 14 days 49.59±4.60; 28 days 62.96±4.68. White matter and gray matter + cerebrospinal fluid (CSF) were determined using a k-means clustering approach, with gray matter and CSF clustered together. Volumes of gray matter + CSF and white matter both increased during the first four postnatal weeks (in cm3:GM+CSF: day 1:37.35±1.04; day 14: 44.86±2.09; day 28: 54.82±2.07; WM: day 1: 2.53±0.74; day 14: 4.73±0.21; day 28: 8.15±0.28). White matter volume increased faster than other clusters during the 28-day period measured, seen in a decreasing gray matter + CSF to white matter ratio (ratios: day 1: 14.80±0.19; day 14: 9.49±0.09; day 28: 6.72±0.06). These results show that white matter volume increases stronger than gray matter and CSF during the first four weeks postnatally. Furthermore, these results provide a baseline measurement for normal brain growth curves in piglets, which can serve as template for future examination of effects of intrauterine growth retardation on brain development.

Use of a competition index to describe differences in physiological parameters associated with energy metabolism and stress in overstocked Holstein dairy cattle

Huzzey, J.M.[1], Nydam, D.V.[1], Grant, R.J.[2] and Overton, T.R.[1], [1]Cornell University, Ithaca, NY, 14850, USA, [2]W H Miner Institute, Chazy, NY, 12921, USA; jmh425@cornell.edu

The objective of this study was to evaluate how physiological parameters associated with stress and energy metabolism are affected based on a cow's ability to compete for limited access to the feed bunk. Forty Holstein dairy cattle were housed in an overstocked pen (5 stalls/10 cows and 0.34 m linear feed bunk (FB) space/cow) in groups of 10 (4 primiparous and 6 multiparous cows) for 14 d. Plasma NEFA, glucose and insulin were measured from blood sampled every 2 d and during a glucose tolerance test (GTT) performed on d 13. Feces, collected every 2 d, were analyzed for fecal cortisol metabolites (FCORT). Plasma cortisol response to an ACTH challenge was measured on d 14. Feeding behavior and displacements at the FB were recorded from d 7 to d 10 of the observation period. A competition index (CI) was calculated for each cow by dividing the number of displacements the animal initiated at the FB by the total number of displacements the animal was involved in, either as an initiator or receiver. Cows were also divided into 3 sub-groups based on their CI: High-ranking (HR: $CI \geq 0.6$), Moderate-ranking (MR: $0.4 \leq CI < 0.6$), and Low-ranking (LR: $CI<0.4$). Primiparous cows accounted for 7%, 36% and 79% of the total number of animals in the HR (n=15), MR (n=11), and LR (n=14) groups, respectively. Considering CI as a continuous variable, total daily feeding time and proportion of total daily time spent at the FB 3 h after fresh feed delivery were positively correlated with CI (r=0.33 and 0.41, respectively; $P \leq 0.03$). CI was negatively correlated with average daily NEFA concentration (r=-0.63, P<0.001), glucose concentration (r=-0.58, P<0.001) and FCORT (r=-0.59, P<0.001), but not correlated with average daily insulin concentration. Previous research has shown that high NEFA is a risk factor for disease therefore LR animals may be at greater risk for future health complications. Higher FCORT in LR animals may indicate a greater stress response among animals that have more difficulty competing for access to feed. During the GTT there was no difference in glucose response curves between the 3 CI categories as measured by area under the curve (AUC) analysis (P=0.24). Insulin production during the GTT was highest among LR cows and lowest among HR cows (AUC=6821 vs. 2765 $\mu IU/l \times 180$ min, P=0.04); this could suggest improved insulin sensitivity in HR animals. Cortisol production in response to ACTH was not different among animals in the 3 CI categories (P=0.53). This competition index can be used to detect physiological differences in cattle responding to an overstocked environment.

Authors index

E

F

G

H

Printed in the United States
by Baker & Taylor Publisher Services